W0082244

Selected Titles in This Series

619 **Kazuyoshi Kiyohara,** Two classes of Riemannian manifolds whose geodesic flows are integrable, 1997

618 **Karl H. Hofmann and Wolfgang A. F. Ruppert,** Lie groups and subsemigroups with surjective exponential function, 1997

617 **Robin Hartshorne,** Families of curves in $\mathbb{P}^3$ and Zeuthen's problem, 1997

616 **Serguei G. Bobkov and Christian Houdré,** Some connections between isoperimetric and Sobolev-type inequalities, 1997

615 **Michael A. Dritschel and Hugo J. Woerdeman,** Model theory and linear extreme points in the numerical radius unit ball, 1997

614 **Richard Warren,** The structure of k-CS-transitive cycle-free partial orders, 1997

613 **D. L. Flannery,** The finite irreducible linear 2-groups of degree 4, 1997

612 **Joan Porti,** Torsion de Reidemeister pour les variétés hyperboliques, 1997

611 **D. Ginzburg, I. Piatetski-Shapiro, and S. Rallis,** L functions for the orthogonal group, 1997

610 **Mark Hovey, John H. Palmieri, and Neil P. Strickland,** Axiomatic stable homotopy theory, 1997

609 **Liviu I. Nicolaescu,** Generalized symplectic geometries and the index of families of elliptic problems, 1997

608 **Christina Q. He and Michel L. Lapidus,** Generalized Minkowski content, spectrum of fractal drums, fractal strings, and the Riemann zeta-functions, 1997

607 **Adele Zucchi,** Operators of class C_0 with spectra in multiply connected regions, 1997

606 **Moshé Flato, Jacques C. H. Simon, and Erik Taflin,** Asymptotic completeness, global existence and the infrared problem for the Maxwell-Dirac equations, 1997

605 **Liangqing Li,** Classification of simple C^*-algebras: Inductive limits of matrix algebras over trees, 1997

604 **Hajnal Andréka, Steven Givant, and István Németi,** Decision problems for equational theories of relation algebras, 1997

603 **Bruce N. Allison, Saeid Azam, Stephen Berman, Yun Gao, and Arturo Pianzola,** Extended affine Lie algebras and their root systems, 1997

602 **Igor Fulman,** Crossed products of von Neumann algebras by equivalence relations and their subalgebras, 1997

601 **Jack E. Graver and Mark E. Watkins,** Locally finite, planar, edge-transitive graphs, 1997

600 **Ambar Sengupta,** Gauge theory on compact surfaces, 1997

599 **Tai-Ping Liu and Yanni Zeng,** Large time behavior of solutions for general quasilinear hyperbolic-parabolic systems of conservation laws, 1997

598 **Valentina Barucci, David E. Dobbs, and Marco Fontana,** Maximality properties in numerical semigroups and applications to one-dimensional analytically irreducible local domains, 1997

597 **Ragnar-Olaf Buchweitz and John J. Millson,** CR-geometry and deformations of isolated singularities, 1997

596 **Paul S. Bourdon and Joel H. Shapiro,** Cyclic phenomena for composition operators, 1997

595 **Eldar Straume,** Compact connected Lie transformation groups on spheres with low cohomogeneity, II, 1997

594 **Solomon Friedberg and Hervé Jacquet,** The fundamental lemma for the Shalika subgroup of $GL(4)$, 1996

593 **Ajit Iqbal Singh,** Completely positive hypergroup actions, 1996

592 **P. Kirk and E. Klassen,** Analytic deformations of the spectrum of a family of Dirac operators on an odd-dimensional manifold with boundary, 1996

591 **Edward Cline, Brian Parshall, and Leonard Scott,** Stratifying endomorphism algebras, 1996

(*Continued in the back of this publication*)

MEMOIRS
of the
American Mathematical Society

Number 619

Two Classes of Riemannian Manifolds Whose Geodesic Flows Are Integrable

Kazuyoshi Kiyohara

November 1997 • Volume 130 • Number 619 (third of 4 numbers) • ISSN 0065-9266

American Mathematical Society
Providence, Rhode Island

1991 *Mathematics Subject Classification.*
Primary 53C22, 58F05;
Secondary 58F07, 70H20.

Library of Congress Cataloging-in-Publication Data

Kiyohara, Kazuyoshi, 1954–
Two classes of Riemannian manifolds whose geodesic flows are integrable / Kazuyoshi Kiyohara.
p. cm. — (Memoirs of the American Mathematical Society, ISSN 0065-9266 ; no. 619)
"November 1997, volume 130, number 619 (third of 4 numbers)."
Includes bibliographical references.
ISBN 0-8218-0640-8
1. Flows (Differentiable dynamical systems) 2. Geodesics (Mathematics) 3. Riemannian manifolds. I. Title. II. Series.
QA3.A57 no. 619
[QA614.82]
510 s—dc21
[516.3′73] 97-30685
CIP

Memoirs of the American Mathematical Society

This journal is devoted entirely to research in pure and applied mathematics.

Subscription information. The 1998 subscription begins with volume 131 and consists of six mailings, each containing one or more numbers. Subscription prices for 1998 are $435 list, $348 institutional member. A late charge of 10% of the subscription price will be imposed on orders received from nonmembers after January 1 of the subscription year. Subscribers outside the United States and India must pay a postage surcharge of $30; subscribers in India must pay a postage surcharge of $43. Expedited delivery to destinations in North America $35; elsewhere $110. Each number may be ordered separately; *please specify number* when ordering an individual number. For prices and titles of recently released numbers, see the New Publications sections of the *Notices of the American Mathematical Society.*

Back number information. For back issues see the *AMS Catalog of Publications.*

Subscriptions and orders should be addressed to the American Mathematical Society, P. O. Box 5904, Boston, MA 02206-5904. *All orders must be accompanied by payment.* Other correspondence should be addressed to Box 6248, Providence, RI 02940-6248.

Memoirs of the American Mathematical Society is published bimonthly (each volume consisting usually of more than one number) by the American Mathematical Society at 201 Charles Street, Providence, RI 02904-2294. Periodicals postage paid at Providence, RI. Postmaster: Send address changes to Memoirs, American Mathematical Society, P. O. Box 6248, Providence, RI 02940-6248.

This publication is indexed in *Science Citation Index*®, *SciSearch*®, *Research Alert*®, *CompuMath Citation Index*®, *Current Contents*®/*Physical, Chemical & Earth Sciences.*
Printed in the United States of America.

∞ The paper used in this book is acid-free and falls within the guidelines established to ensure permanence and durability.
Visit the AMS home page at URL: `http://www.ams.org/`

10 9 8 7 6 5 4 3 2 1   02 01 00 99 98 97

Contents

ABSTRACT. In Part 1 we study global structure of riemannian manifolds whose geodesic flows are integrable with a set of first integrals in involution that are simultaneously normalizable quadratic forms on each cotangent space. We call them *Liouville manifolds*. After developing a general theory, we classify the isomorphism classes of "proper Liouville manifolds of rank one". As an application, a new family of $C_{2\pi}$-metrics on the sphere is given.

In Part 2 we study n-dimensional Kähler manifolds whose geodesic flows possess n first integrals in involution that are fibrewise hermitian forms and simultaneously normalizable. Under some mild assumption, one can associate with such a manifold an n-dimensional commutative Lie algebra of infinitesimal automorphisms. This, combined with the given n first integrals, makes the geodesic flow integrable. If the manifold is compact, then it becomes a toric variety.

Key words and phrases. Integrable geodesic flow, Liouville manifold, Integrable hamiltonian system, Ellipsoid, Toric variety, Kähler manifold.

Preface

This paper contains two parts. In each of them we deal with a particular class of riemannian manifolds whose geodesic flows are integrable. They are called Liouville manifolds and Kähler-Liouville manifolds respectively. Each part can be read separately except Section 7 of Part 2, where the main results in Part 1 are necessary.

Kazuyoshi Kiyohara

Part 1. Liouville Manifolds

Introduction

Since Jacobi it is well known that the geodesic flows of ellipsoids are integrable in the sense of symplectic geometry (or in Liouville's sense). In fact they possess n (= the dimension of the ellipsoid) first integrals in involution that are quadratic forms on each cotangent space. Moreover those first integrals have a remarkable property; they are simultaneously normalizable on each fibre. If one uses the so-called elliptic coordinate system to describe the geodesic equation, then those first integrals are expressed in normalized form (and the variables are separated).

After Jacobi's work, Liouville obtained a general form of integrable hamiltonian systems that are integrated in a similar way as the geodesic flows of ellipsoids, which we refer to as Liouville's system (see [**L**] pp. 700–708). Liouville's system (the case without potential) is expressed as follows: Let $(x_1, \ldots, x_n, \xi_1, \ldots, \xi_n)$ be a canonical coordinate system, and let $b_{ij}(x_i)$ be n^2 functions in one variable such that $\det(b_{ij}) \neq 0$. Then the functions F_i defined by

$$\sum_j b_{ij} F_j = \xi_i^2$$

satisfies $\{F_i, F_j\} = 0$. (cf. Proposition 1.1.3.) Here, for example, F_1 is the hamiltonian, and $F_1, \ldots, F_n$ are first integrals of the system.

Later, Stäckel characterized Liouville's system in terms of the properties of the first integrals. Namely, let $F_1 = H, F_2, \ldots, F_n$ be a set of functions in involution on the cotangent bundle of n-dimensional manifold. Assume that every F_i is a quadratic form, and those quadratic forms are linearly independent and simultaneously normalizable on each cotangent space. Then one can find a coordinate system with which the hamiltonian system defined by the hamiltonian H is described as Liouville's system (cf. Klingenberg [**Kl1**] and Proposition 1.1.3).

The main purpose of this part is to investigate global properties of complete riemannian manifolds whose geodesic flows are regarded as Liouville's system (at least on a dense subset of the cotangent bundle). We call the pair of such a riemannian manifold and the space of first integrals *Liouville manifold.* The precise definition is as follows: Let M be an n-dimensional complete riemannian manifold, and let E be the associated energy function (the hamiltonian of the geodesic flow). Let $\mathcal{F}$ be an n-dimensional vector space of functions on the cotangent bundle T^*M that are fibrewise homogeneous polynomials of degree two. We say that the pair $(M, \mathcal{F})$ is a Liouville manifold if the following four conditions are satisfied:

(L.1) $E \in \mathcal{F}$;
(L.2) $\{F, H\} = 0$ for any $F, H \in \mathcal{F}$;
(L.3) $F_p \quad (F \in \mathcal{F})$ are simultaneously normalizable for any $p \in M$;

(L.4) $\dim \{F_p \mid F \in \mathcal{F}\} = n$ at some point $p \in M$;
where $F_p = F|_{T_p^* M}$.

Besides the quadratic surfaces, one can find concrete examples of Liouville manifold defined over the sphere in Brailov's paper [**Br**] (see also [**MF1**], [**MF2**], [**Fo**], [**Mi**], [**Th**], [**IW**] for integrable hamiltonian systems on homogeneous spaces). Also, Yamato [**Y**] recently showed that for some Liouville manifolds a conformal change of the metric yields another Liouville manifold. 2-dimensional Liouville manifolds (Liouville surfaces) were completely investigated by Kolokol'tsov [**Kol1**] (and independently by the author [**Ki1**]) for compact case, and by Sugahara, Igarashi, and the author [**IKS**] for noncompact case (see also [**V**]). There a classification of the isomorphism classes of Liouville surfaces is given in each case.

Now, let $(M, \mathcal{F})$ be a Liouville manifold. Let M^s be the set of points $p \in M$ where the dimension of the vector space

$$\{F_p \mid F \in \mathcal{F}\}$$

is less than n $(= \dim M)$. Put $M^0 = M - M^s$. Since the description by Liouville and Stäckel is valid only on M^0, we encounter the following problems: (1) What kind of subset M^s is; (2) How the coordinate systems of Liouville and Stäckel behave around the subset M^s; (3) How those coordinate systems are globally continued. Also we can ask a fundamental question: (4) What kind of manifolds admit the structure of Liouville manifold?

In the case of the ellipsoid $M : \sum_{i=0}^n y_i^2/a_i = 1$ $(a_0 > \cdots > a_n > 0)$, M^s is the union of the submanifolds I_m $(1 \le m \le n-1)$ of codimension two defined by

$$y_m = 0, \quad \sum_{i \neq m} \frac{y_i^2}{a_i - a_m} = 0.$$

The answer to the questions (2), (3) in this case is given by the existence of the 2^{n-1}-fold branched normal covering

$$\Phi : R = \prod_{i=1}^n (R/\alpha_i Z) \to M$$

whose branch locus is M^s, and whose covering transformation group is generated by

$$\sigma_i(x_1, \ldots, x_n) = (x_1, \ldots, x_{i-1}, -x_i, \frac{\alpha_{i+1}}{2} - x_{i+1}, x_{i+2}, \ldots, x_n) \quad (1 \le i \le n-1),$$

where (x) is the natural coordinate system on R. Also this branched covering possesses the property that (x), regarded as a local coordinate system of M^0, is a variant of the elliptic coordinate system, i.e., the coordinate system of Liouville and Stäckel. (cf. [**Kl2**] pp. 305–306 for 2-dimensional ellipsoids).

We say that a Liouville manifold $(M, \mathcal{F})$ is *proper* if for any $p \in M^s$ and any $F \in \mathcal{F} - \{0\}$ satisfying $F_p = 0$ there is some $\lambda \in T_p^* M$ such that $dF \neq 0$ at λ. In this paper we shall consider proper Liouville manifolds only. An advantage of assuming this condition is that it assures the "non-degeneracy" of the subset M^s (Proposition 1.2.8), which is crucial in determining the structure of M^s. Up to now we do not know exactly how severe this condition is. In the case of Liouville surfaces defined over 2-sphere, proper ones are generic and others (surfaces of revolution) are obtained as a limit of proper ones (cf. [**Kol1**], [**Ki1**]). In general, a Liouville manifold $(M, \mathcal{F})$ is not proper if $\mathcal{F}$ contains the square of a Killing vector field whose zero set is not empty. However, it should be noted that this condition does not imply the non-existence of Killing vector fields on proper Liouville manifolds. For example, the sphere of constant curvature admits the structure of proper Liouville manifold (see Appendix).

We now explain various results in Part 1. Let $(M, \mathcal{F})$ be an n-dimensional proper Liouville manifold. Then our first result is:

1 (COROLLARY 1.3.6). *M^s is a (locally finite) union of closed submanifolds of codimension two. At each intersection point of those submanifolds their normal spaces are mutually orthogonal.*

To each connected, closed submanifold of codimension two contained in M^s, we can associate a unique element $F \in \mathcal{F}$ up to constant factor so that the submanifold is equal with one of the connected components of

$$\{p \in M \mid F_p = 0\}$$

Furthermore, we have the following result.

2 (THEOREM 2.3.3, LEMMA 2.1.8). *There are elements $F_1, \ldots, F_{n-r}$ in $\mathcal{F}$ possessing the following properties:*
(1) *$I_m = \{p \in M \mid (F_m)_p = 0\}$ is a closed submanifold of codimension two;*
(2) *$M^s = \cup_{m=1}^{n-r} I_m$;*
(3) *$F_1, \ldots, F_{n-r}, E$ are linearly independent in $\mathcal{F}$;*
(4) *If $F \in \mathcal{F} - \{0\}$ vanishes at a point $p \in M^s$, then F is equal to $\sum_{m=1}^{n-r} a_m F_m$ for some constants a_m; moreover, if $a_m \neq 0$, then $p \in I_m$ for any m.*

The number r that appears in the statement above is called the *rank* of the proper Liouville manifold, which is an integer between 1 and n. Note that the rank is n if and only if $M^s = \emptyset$. We also have:

3 (PROPOSITIONS 2.1.1, 2.1.9, COROLLARY 2.3.4). *For any $m = 1, \ldots, n-r$,*

$$J_m = \{p \in M \mid \text{the rank of } (F_m)_p \leq 1\}$$

is a connected, closed, totally geodesic submanifold of codimension one. With the induced metric and the induced first integrals J_m becomes a proper Liouville manifold of rank r (= the rank of $(M, \mathcal{F})$).

In Proposition 1.3.7 we give an analogous result to Stäckel's theorem for points on M^s. In particular, we have the following result, which gives an answer to the question (2) stated above.

4. *Let B be the unit disk $|z| < 1$ in $\boldsymbol{C}$, and I the interval $|x| < 1$ in $\boldsymbol{R}$. Let p_0 be a point in $\cap_{j=1}^{k} I_{m_j}$ such that $p_0 \notin I_m$ for $m \neq m_j$ $(1 \leq j \leq k)$. Then there is a neighborhood U of p_0 diffeomorphic to $B^k \times I^{n-2k}$ such that the branched covering*

$$\phi : B^k \times I^{n-2k} \to U$$

defined by

$$\phi(z_1, \ldots, z_k, x_{2k+1}, \ldots, x_n) = (z_1^2, \ldots, z_k^2, x_{2k+1}, \ldots, x_n)$$

possesses the following properties:
(1) *$U \cap I_{m_j} = \{\phi(z, x) \in U \mid z_j = 0\}$;*

(2) *by putting* $z_i = x_{2i-1}+\sqrt{-1}x_{2i}$, $(x_1, \ldots, x_n)$ *defines an admissible coordinate system through* ϕ *on a neighborhood of each point in* $U \cap M^0$.

Here the term"admissible coordinate system" stands for the same thing as coordinate system of Liouville and Stäckel, i.e., a coordinate system that normalizes elements of $\mathcal{F}$ simultaneously. Let $p_0 \in M^s$ and (z, x) be as above. Then we have k 2-dimensional subspaces $N_{p_0}I_{m_j}$ (the normal space to I_{m_j}) and $n-2k$ 1-dimensional subspaces spanned by $\partial/\partial x_i$ $(2k+1 \leq i \leq n)$ of $T_{p_0}M$. We say that a subspace W of $T_{p_0}M$ is *admissible* if it is a sum of those subspaces. For a point $p \in M^0$ we say that a subspace W of T_pM is admissible if it is spanned by a subset of $\{\partial/\partial x_j \mid 1 \leq j \leq n\}$, where (x) is an admissible coordinate system around p. A submanifold N is said to be *admissible* if T_pN is admissible for every $p \in N$. The following theorem gives a partial answer to the continuation problem.

5 (THEOREM 2.2.3). *Let* $p \in M^0$ *and* W *be an admissible subspace of* T_pM. *In case* $p \in J_m$, *we also assume that* $W \not\subset T_pJ_m$. *Then there is a unique connected admissible submanifold* N *of* M *passing through* p *such that* $T_pN = W$ *and* N *is complete with respect to the induced metric. Moreover, with the induced first integrals* N *becomes a proper Liouville manifold such that* $N^s = N \cap M^s$.

Up to now, further results are obtained only for proper Liouville manifolds of rank one. To state those we first explain the notion of the core of a proper Liouville manifold. Let $(M, \mathcal{F})$ be a proper Liouville manifold of rank r, and put $C = \cap J_m$. Then C becomes a connected, proper Liouville manifold with the dimension and the rank being r by restricting the metric and the first integrals to C, which we call the *core submanifold* (cf. Section 2.3). Since the rank of F_m $(1 \leq m \leq n-r)$ at points on C is 1 or 0, we obtain a smooth function f_m on C so that F_m is described as $f_m V_m^2$ there for some unit vector V_m. Let $[f_m]$ be the equivalence class obtained by ignoring the non-zero constant multiplication. Then we call the pair

$$(C, \{[f_1], \ldots, [f_{n-r}]\})$$

of the core submanifold and the set of the $n-r$ equivalence classes of functions on it the *core* of the Liouville manifold. In the case that the rank of the proper Liouville manifold is one, the core becomes a pair of a 1-dimensional riemannian manifold and the $n-1$ equivalence classes of functions on it.

Now, let us assume that the rank is one. Then using the results in the 2-dimensional case, we see that the cores, and therefore the Liouville manifolds themselves are roughly classified into 4 types; (A), (B), (C), and (D). (Ellipsoids are of type (A).) In each case we investigate the conditions that the $n-1$ functions contained in the cores must satisfy, and obtain the notion of *possible cores.* By definition, two Liouville manifolds $(M, \mathcal{F})$ and $(M', \mathcal{F}')$ are said to be isomorphic if there is an isometry from M to M' that also maps $\mathcal{F}$ to $\mathcal{F}'$. Then we have the following result.

6 (THEOREM 3.4.1). *There is a one-to-one correspondence between the isomorphism classes of proper Liouville manifolds of rank one and the isomorphism classes of possible cores.*

For the precise description of possible cores and their isomorphisms, see Section 3.2. Here we only mention the fact that the isomorphism classes of possible cores are still parametrized by $n-1$ functions in single variable. We also obtain the following results.

7 (THEOREMS 3.3.1, 3.4.1). *There is a 2^{n-1}-fold branched normal covering $\Phi : R \to M$ possessing the similar properties as the case of ellipsoids (stated before) if the type of $(M, \mathcal{F})$ is* (A) *or* (C) *or* (D). *R is the product of n circles if the type is* (A), *the product of $n-1$ circles and a line if the type is* (C), *and the product of $n-2$ circles and 2 lines if the type is* (D).

8 (THEOREMS 3.3.1, 3.4.1). *M is diffeomorphic to the n-sphere S^n if the type of $(M, \mathcal{F})$ is* (A), *to the real projective space $\boldsymbol{R}P^n$ if the type is* (B), *and to $\boldsymbol{R}^n$ if the type is* (C) *or* (D).

It also turns out that the (unbranched) double covering of a Liouville manifold of type (B) becomes of type (A). In the case of flat $\boldsymbol{R}^n$, the coordinate system defined through Φ is essentially the same as the classical elliptic (resp. parabolic) coordinate system if the type is (C) (resp. (D)) (cf. Appendix).

Next let us consider the following problem: To what extent does the underlying riemannian manifold determine the structure of a proper Liouville manifold? To this problem we obtain the following partial answer.

9 (THEOREM 3.5.2). *Let $(M, \mathcal{F})$ be a proper Liouville manifold of rank one and type* (A). *Assume that none of the $n-1$ functions contained in the core has the form $a\cos^2 2\pi t/l + b$. Then $\mathcal{F}$ is unique. Here a and b are constants, and l is the length of the core submanifold (a circle in this case). Namely, if $(M, \mathcal{F}')$ is another proper Liouville manifold, then $\mathcal{F}' = \mathcal{F}$.*

In the case of proper Liouville manifolds of rank one defined over the sphere of constant curvature 1, all of the $n-1$ functions in the core is of the form $a\cos^2 t + b$ (cf. Proposition A.1.1). It would be worth noted that the conclusion of the result above is a little bit stronger than usually expected, which would be "$(M, \mathcal{F}')$ is isomorphic to $(M, \mathcal{F})$". This fact enables us to get the following result.

10 (COROLLARY 3.5.4). *Under the same assumptions as those of the result 9, the isometry group of M is finite.*

As an application of the results above, we obtain the following

11 (THEOREM 3.6.1). *Among proper Liouville manifolds of rank one and type* (A), *there is a family of $C_{2\pi}$-manifolds, any of which admits no non-trivial Killing vector field except for the sphere of constant curvature* 1.

By definition, a riemannian manifold is called a $C_{2\pi}$-manifold, and the metric is called a $C_{2\pi}$-metric if all of the geodesics are closed, and have the common length 2π. Various results and references on this subject can be found in Besse **[Be]** (see also **[Kol2]**, **[Ki1]**, **[Ki2]**, **[Ki3]**, **[Ts1]**, **[Ts2]** for later progress). As far as the author knows, Weinstein's $C_{2\pi}$-manifolds described in **[Be]** p. 120 have been the only non-standard examples in dimension greater than 2. Since those $C_{2\pi}$-manifolds admit non-trivial Killing vector fields, it follows that they are different from ours. It is an interesting fact that Weinstein's $C_{2\pi}$-manifolds are also regarded as (non-proper) Liouville manifolds, which one can easily verify in **[Be]** p. 120.

In Appendix we determine all of the proper Liouville manifolds of rank one defined over simply connected manifolds of constant curvature. In particular, we have:

12 (APPENDIX). *Let $(M, \mathcal{F})$ be a proper Liouville manifold of rank one such that M is simply connected and of constant curvature. Then $\mathcal{F}$ is a subspace of the*

symmetric tensor product $\mathcal{S}^2\mathfrak{g}$, *where* $\mathfrak{g}$ *denotes the Lie algebra of Killing vector fields.*

We now briefly explain the organization of this part. In Sections 1.2 and 1.3 we shall investigate the local structure of proper Liouville manifolds around a point in M^s. The main results here are 1 and 4 described above. In Sections 2.1, 2.2, and 2.3 we shall investigate the global structure of proper Liouville manifolds. Here we study the properties of the submanifolds J_m, with which we obtain the definition of the rank and the core. The properties of admissible submanifolds are also studied. The results 2, 3, and 5 are proved. Through Sections 3.1–3.6 we shall concentrate our attention on proper Liouville manifolds of rank one. The main ingredient here is the classification theory for the isomorphism classes of those manifolds, which will be established through Sections 3.1–3.4. Sections 3.5 and 3.6 contain the results 9, 10, and 11 stated above.

In Appendix we shall determine all the proper Liouville manifolds of rank one whose underlying riemannian manifolds are simply connected and of constant curvature. First we shall determine those cores, and then give the corresponding spaces $\mathcal{F}$ of first integrals as subspaces of $\mathcal{S}^2\mathfrak{g}$, where $\mathfrak{g}$ is the Lie algebra of Killing vector fields. Here we also give the branched coverings stated in the result 7 in explicit forms, which is necessary in proving the result 8. Although part of this appendix is well known (e.g. see Moser [**Mo**]), we have decided to include it in this appendix for convenience.

The author is grateful to Professor Minoru Tanaka, who showed the author that the former half of Proposition 1.1.3 is due to Stäckel, and also recommended him Lützen's book [**L**].

Preliminary remarks and notations

Let M be a manifold, and let $\pi : T^*M \to M$ be the bundle projection of the cotangent bundle. The canonical 1-form α on T^*M is defined by the formula:

$$\alpha(X) = \lambda(\pi_* X), \qquad X \in T_\lambda(T^*M).$$

As usual, we regard T^*M as a symplectic manifold with respect to the symplectic 2-form $d\alpha$. For a function f on T^*M we denote by X_f the symplectic vector field defined by the "hamiltonian" f, i.e.,

$$i_{X_f} d\alpha = -df,$$

where i denotes the inner derivation. If h is another function, then the Poisson bracket $\{f, h\}$ is given by $X_f h$.

Let u be a function on T^*M that is fibrewise homogeneous polynomials of degree k. Then u is naturally identified with a section of the vector bundle $\mathcal{S}^kTM$, the symmetric tensor product of k copies of the tangent bundle TM, by using the duality of TM and T^*M. In this paper we shall often identify them without any comment. Note that if u and v are those of degree one, then the Poisson bracket and the Lie bracket of them coincide under this identification.

Now, let M be a riemannian manifold. The *energy function* E on T^*M is defined as a half of the square of the norm function. The geodesic flow of M is then given as a symplectic flow on T^*M generated by X_E. The geodesic flow of M is called *integrable* if there are n functions $F_1, \dots, F_n = E$ on T^*M − {0-section} such that $\{F_i, F_j\} = 0$ for any i, j, and $dF_1 \wedge \cdots \wedge dF_n \neq 0$ almost everywhere.

Let F be a function on T^*M that is fibrewise homogeneous polynomial of degree two. Then we denote by F_p ($p \in M$) the quadratic form on T_p^*M obtained by restricting F to T_p^*M. We also define the self-adjoint endomorphism F_p^e of T_p^*M by the formula:

$$g(F_p^e(\lambda), \lambda) = F(\lambda), \qquad \lambda \in T_p^*M,$$

where g denotes the fibre metric of T^*M. Note that $2E_p^e$ is the identity.

The riemannian structure on M induces the bundle isomorphism $T^*M \to TM$ in the usual way, which we shall denote by $\sharp$. Let N be a submanifold of M. Then we also have the isomorphism $\sharp : T^*N \to TN$ with respect to the induced metric. Composing those isomorphisms and the differential of the inclusion $N \to M$, we obtain an injective mapping $T^*N \to T^*M$. With this mapping we regard T^*N as a submanifold of T^*M. It is easily seen that the pull-back image of the canonical 1-form on T^*M by this mapping coincides with that on T^*N.

In Part 1 the symbol $C^\infty(X)$ always represents the vector space of C^∞ functions on the manifold X, even if X has a bundle structure. For the space of sections of a vector bundle Y over a manifold, we shall use the symbol $\Gamma(Y)$.

The unit sphere in the tangent space (resp. in the cotangent space) at a point $p \in M$ is denoted by S_pM (resp. S_p^*M).

1. Local Structure of Proper Liouville Manifolds

1.1. Liouville manifolds and the properness

Let $(M, \mathcal{F})$ be an n-dimensional Liouville manifold. We put

$$\mathcal{F}_p = \{F_p \mid F \in \mathcal{F}\}, \qquad p \in M.$$

By definition, M^0 is the set of points $p \in M$ such that the dimension of $\mathcal{F}_p$ is equal to n.

Proposition 1.1.1. *M^0 is open and dense in M.*

Proof. The condition (L.4) in the definition of Liouville manifold stated in Introduction implies that M^0 is not empty. Let $p_0 \in M^0$. Let $x = (x_1, \dots, x_n)$ be a local coordinate system around p_0 so that $x(p_0) = 0$, and let (x, ξ) be the associated canonical coordinate system on the cotangent bundle. By virtue of the condition (L.3) we can take x and a basis $F_1, \dots, F_n$ of $\mathcal{F}$ so that

$$F_k = \sum_{ij} a_k^{ij}(x)\xi_i\xi_j, \quad a_k^{ij}(0) = \begin{cases} 1 & \text{if } \ i = j = k, \\ 0 & \text{otherwise.} \end{cases}$$

Then we get

$$dF_1 \wedge \cdots \wedge dF_n \wedge dx_1 \wedge \cdots \wedge dx_n = 2^n \xi_1 \dots \xi_n \, d\xi_1 \wedge \cdots \wedge d\xi_n \wedge dx_1 \wedge \cdots \wedge dx_n$$

at $(0, \xi)$. This implies that the set U of unit covectors at p_0 where $dF_1 \wedge \cdots \wedge dF_n \neq 0$ is open and dense in $S_{p_0}^*M$.

Now take $\lambda \in U$. Let $\pi : T^*M \to M$ be the bundle projection, and let $\{\zeta_t\}$ be the geodesic flow. Then it is easy to see that the Jacobi fields

$$Y_i(t) = \pi_*(\zeta_t)_*(X_{F_i})_\lambda \qquad (1 \le i \le n)$$

along the geodesic $\gamma(t) = \pi(\zeta_t\lambda)$ are linearly independent. Moreover, since X_{F_1}, $\dots, X_{F_n}$ form a lagrangean subspace, it follows that

$$g(Y_i(t), \nabla_{\partial/\partial t}Y_j(t)) = g(\nabla_{\partial/\partial t}Y_i(t), Y_j(t)).$$

for any i, j (cf. [**Be**] p. 46). Now suppose that $Y_1(t_0), \dots, Y_n(t_0)$ are not linearly independent at some t_0. Replacing Y_i with their linear combinations, we may assume that $Y_1(t_0), \dots, Y_k(t_0)$ are linearly independent and $Y_{k+1}(t_0) = \cdots = Y_n(t_0) = 0$. Then the property above implies that $Y_1(t_0), \dots, Y_k(t_0), \nabla_{\partial/\partial t}Y_{k+1}(t_0)$, $\dots, \nabla_{\partial/\partial t}Y_n(t_0)$ are linearly independent. Hence we see that the set A_λ of time t such that $Y_1(t), \dots, Y_n(t)$ does not form a basis of $T_{\gamma(t)}M$ is discrete in $\boldsymbol{R}$.

Since $\pi(\zeta_t\lambda) \in M^0$ for $\lambda \in U$ and $t \in \boldsymbol{R} - A_\lambda$, it follows that M^0 is a dense subset of M. The openness is clear. □

COROLLARY 1.1.2. *Let $F_1, \dots, F_n$ be a basis of $\mathcal{F}$. Then*

$$dF_1 \wedge \cdots \wedge dF_n \neq 0$$

on an open and dense subset of T^*M. In particular, the geodesic flow of the riemannian manifold M is integrable.

PROOF. As shown in the proof of Proposition 1.1.1, $dF_1 \wedge \cdots \wedge dF_n$ does not vanish on an open and dense subset of T_p^*M, provided $p \in M^0$. Hence the corollary follows from the proposition. □

The following proposition is due to Liouville and Stäckel (see Lützen [**L**] p. 703, p. 705 and Klingenberg [**Kl1**] pp. 183–185).

PROPOSITION 1.1.3. *Let $F_1, \dots, F_n$ be a basis of $\mathcal{F}$. Then for any point $p \in M^0$ there are a neighborhood U of p, a coordinate system $x = (x_1, \dots, x_n)$ on U, and an $n \times n$ matrix-valued function $B = (b_{ij})$ on U such that b_{ij} depends only on the variable x_i, and*

$$\sum_j b_{ij}F_j = \xi_i^2 \qquad (1 \le i \le n),$$

where (x, ξ) denotes the canonical coordinate system on T^*U associated with x. Moreover, n line subbundles of TU spanned by each $\partial/\partial x_i$ are uniquely determined by $\mathcal{F}$ except the ordering. Conversely, for a given $B = (b_{ij}(x_i))$ with $\det B \neq 0$, the vector space $\mathcal{F}$ spanned by the functions F_i defined by the formula above satisfies the conditions (L.2), (L.3), *and* (L.4).

PROOF. Let $p \in M^0$. By virtue of the condition (L.3) in the definition of Liouville manifold we can take an orthonormal frame $V_1, \dots, V_n$ around p so that each F_i is written as $\sum_j f_{ij}V_j^2$, where f_{ij} are C^∞ functions on a neighborhood of p. Clearly this orthonormal frame is uniquely determined by the metric g and $\mathcal{F}$ except for the sign and the order. Let (a_{ij}) be the inverse matrix of (f_{ij}). Then

$$\begin{aligned} \{V_k^2, V_l^2\} &= \{\sum_i a_{ki}F_i, \sum_j a_{lj}F_j\} \\ &= \sum_{ij}(\{F_i, a_{lj}\}a_{ki}F_j + \{a_{ki}, F_j\}a_{lj}F_i) \qquad (1.1.1) \\ &= \sum_j \{V_k^2, a_{lj}\}F_j + \sum_i \{a_{ki}, V_l^2\}F_i. \end{aligned}$$

Note that each formula in the equalities above is fibrewise homogeneous polynomial of degree three in the variables $V_1, \ldots, V_n$. The first one being divisible by $V_k V_l$, it follows that the quadratic form $\sum_i \{a_{ki}, V_l\} F_i$ is divisible by V_k, provided $k \neq l$. Since each F_i is fibrewise a linear combination of $V_1^2, \ldots, V_n^2$, we have

$$\sum_i \{a_{ki}, V_l\} F_i = c_{kl} V_k^2 \qquad (k \neq l), \tag{1.1.2}$$

c_{kl} being a function on a neighborhood of p. Since

$$V_k^2 = \sum_i a_{ki} F_i,$$

we then have

$$\{a_{ki}, V_l\} = c_{kl} a_{ki} \qquad (k \neq l). \tag{1.1.3}$$

Also, by (1.1.1) and (1.1.2) we get

$$2\{V_k, V_l\} = c_{kl} V_k - c_{lk} V_l \qquad (k \neq l). \tag{1.1.4}$$

For each k, choose a_{ki} that does not vanish at p, and put $A_k = |a_{ki}|$. Put

$$b_{ki} = a_{ki}/A_k, \quad X_k = V_k/\sqrt{A_k} \qquad (1 \leq k, i \leq n).$$

Then from (1.1.3) and (1.1.4) one can easily get

$$\{b_{ki}, X_l\} = 0 \quad (k \neq l), \qquad \{X_k, X_l\} = 0.$$

Regarding $X_1, \ldots, X_n$ as vector fields on a neighborhood of p, we obtain a coordinate system $(x_1, \ldots, x_n)$ around p so that $X_k = \partial/\partial x_k$, and $(\partial/\partial x_l) b_{ki} = 0$ $(k \neq l)$. This completes the proof of the former half of the proposition. The latter half is also seen by a similar computation. □

A Liouville manifold $(M, \mathcal{F})$ is called *proper* if for any $F \in \mathcal{F} - \{0\}$ and $p \in M$ satisfying $F_p = 0$, the vector field X_F (or equivalently, the 1-form dF) does not vanish at some $\lambda \in T_p^* M$. *In the rest of Part 1 we shall always assume that Liouville manifolds are proper.*

1.2. Infinitesimal structure at a point in M^s

Let $(M, \mathcal{F})$ be a proper Liouville manifold. We take a point $p_0 \in M^s$ and fix it throughout Sections 1.2 and 1.3. Let $\dim \mathcal{F}_{p_0} = n - k$ $(1 \leq k \leq n-1)$, and put

$$W = T_{p_0}^* M.$$

If $F \in \mathcal{F} - \{0\}$ satisfies $F_{p_0} = 0$, then it is easily seen that the vector field X_F is tangent to the fibre W. We denote by $X_F|_W$ the restriction of X_F to W. The vector field $X_F|_W$ is described as

$$\sum_{ijk} a_{ij}^k w_i w_j \frac{\partial}{\partial w_k}, \qquad a_{ij}^k \in \mathbf{R}$$

in terms of a linear coordinate system (w_i) on W. Hence $X_F|_W$ is naturally regarded as an element of $S^2 W^* \otimes W$, where $S^2 W^*$ denotes the symmetric tensor product of two copies of the dual space W^* of W.

Let $F_1, \dots, F_n$ be a basis of $\mathcal{F}$ so that $(F_i)_{p_0} = 0$ for $1 \le i \le k$. We may assume that $F_n = 2E$.

LEMMA 1.2.1. (1) *The k vectors $X_{F_1}|_W, \dots, X_{F_k}|_W$ of $\mathcal{S}^2W^* \otimes W$ are linearly independent.*

(2) *As vector fields on W, $X_{F_i}|_W$ $(1 \le i \le k)$ are mutually commutative and contained in the kernel of dF_j $(k+1 \le j \le n)$. Especially they are tangent to the unit sphere $S^*_{p_0}M$ in W.*

PROOF. (1) If they are linearly dependent, then there is $F = \sum_{i=1}^k a_i F_i$ $(\neq 0,$ $a_i \in \boldsymbol{R})$ such that $X_F|_W = 0$. But this contradicts the properness of the Liouville manifold $(M, \mathcal{F})$.

(2) Since $\{F_i, F_j\} = 0$, it follows that $X_{F_i}F_j = 0$ and $[X_{F_i}, X_{F_j}] = 0$. Then the assertion follows by restricting them to W. The last assertion is clear, because $S^*_{p_0}M$ is defined by $F_n = 2E = 1$ in W. □

Now, let us consider the symmetric endomorphisms $F^e_{k+1}, \dots, F^e_n$ of W. Since they are mutually commutative, it follows that W is decomposed to the simultaneous eigenspaces

$$W = W_1 \oplus \cdots \oplus W_l. \tag{1.2.1}$$

Here $F^e_j|_{W_r}$ are scalar multiplications, and there is $F = \sum_{j=k+1}^n a_j F_j$ $(a_j \in \boldsymbol{R})$ such that the eigenvalues of F^e on W_r $(1 \le r \le l)$ are mutually distinct.

LEMMA 1.2.2. $l \ge n - k$.

PROOF. Let $Q_j : W \to W_j$ be the projection with respect to the decomposition (1.2.1). Then it is clear that the subspace of $\mathrm{End}(W)$ spanned by $F^e_{k+1}, \dots, F^e_n$ is contained in the subspace spanned by $Q_1, \dots, Q_l$, which shows the lemma. □

Put $\dim W_r = i_{r+1} - i_r$ $(0 = i_1 < i_2 < \cdots < i_{l+1} = n)$. Let $W^* (\simeq T_{p_0}M)$ be the dual space to W, and let $W^*_1, \dots, W^*_l$ be the subspaces of W^* dual to $W_1, \dots,$ W_l. Take an orthonormal basis $V_1, \dots, V_n$ of W^* so that W^*_r is spanned by $V_{i_r+1},$ $\dots, V_{i_{r+1}}$. The following lemma is obvious.

LEMMA 1.2.3. *There are a neighborhood U of p_0 and unique subbundles $\overline{W}_1,$ $\dots, \overline{W}_l$ of T^*U satisfying the following conditions:*

(1) $(\overline{W}_j)_{p_0} = W_j$*;*
(2) $T^*U = \sum_{j=1}^l \overline{W}_j$ (direct sum)*;*
(3) *Each $(F_i)^e_p$ $(k+1 \le i \le n)$ preserves the subspaces $(\overline{W}_j)_p$ for any $p \in U$;*
(4) *F^e_p has no common eigenvalues on $(\overline{W}_i)_p$ and $(\overline{W}_j)_p$ if $i \neq j$, $p \in U$, where $F = \sum_{j=k+1}^n a_j F_j$ is the one stated before Lemma 1.2.2.*

Let $\overline{W}^*_1, \dots, \overline{W}^*_l$ be the subbundles of TU dual to $\overline{W}_1, \dots, \overline{W}_l$.

LEMMA 1.2.4. *Let $\Gamma(\overline{W}^*_j)$ denote the vector space of sections of $\overline{W}^*_j$. Then*

$$[\Gamma(\overline{W}^*_i), \Gamma(\overline{W}^*_j)] \subset \Gamma(\overline{W}^*_i + \overline{W}^*_j)$$

for any i, j.

PROOF. Since $(\overline{W}_j)_p$ are sum of eigenspaces for $p \in U$, we have

$$[\Gamma(\overline{W}_i^*), \Gamma(\overline{W}_j^*)] \subset \Gamma(\overline{W}_i^* + \overline{W}_j^*)$$

on $U \cap M^0$ by virtue of the formula (1.1.4). Since M^0 is open and dense in M, the lemma follows. □

We now extend the basis $V_1, \ldots, V_n$ to an orthonormal frame on U so that $\overline{W}_r^*$ is spanned by $V_{i_r+1}, \ldots, V_{i_{r+1}}$ at each point on U for any r. Clearly each $(F_i)_p^e$ $(1 \leq i \leq k)$ also preserves the subspaces $(W_r)_p$ because of the property (4) in Lemma 1.2.3. Hence we can describe F_i $(1 \leq i \leq k)$ as

$$(F_i)_p = \sum_{m=1}^{l} \sum_{r,s=i_m+1}^{i_{m+1}} f_{rs}(p) V_r V_s, \qquad p \in U, \quad f_{rs} = f_{sr} \in C^\infty(U).$$

LEMMA 1.2.5. *$\{V_t, f_{rs}\} = 0$ at p_0 if $m \neq m'$, $i_{m'} + 1 \leq t \leq i_{m'+1}$, and $i_m + 1 \leq r, s \leq i_{m+1}$.*

PROOF. In the formula

$$0 = \{2E, F_i\},$$

observe the terms that are linear in the variables V_t, $i_{m'} + 1 \leq t \leq i_{m'+1}$, and quadratic in the variables V_r, $i_m + 1 \leq r \leq i_{m+1}$. First we have

$$0 = \left\{ \sum_{t=i_{m'}+1}^{i_{m'+1}} V_t^2, \sum_{r,s=i_m+1}^{i_{m+1}} f_{rs} V_r V_s \right\} + \left\{ \sum_{j=i_m+1}^{i_{m+1}} V_j^2, \sum_{r,s=i_{m'}+1}^{i_{m'+1}} f_{rs} V_r V_s \right\}.$$

Since $f_{rs}(p_0) = 0$,

$$0 = \sum_{t=i_{m'}+1}^{i_{m'+1}} \sum_{r,s=i_m+1}^{i_{m+1}} \{V_t, f_{rs}\} V_t V_r V_s$$

at p_0, which indicates the lemma. □

PROPOSITION 1.2.6. *$X_{F_i}|_W \in \mathcal{S}^2 W^* \otimes W$ belongs to the subspace*

$$\sum_{r=1}^{l} \mathcal{S}^2 W_r^* \otimes W_r$$

for any i $(1 \leq i \leq k)$.

PROOF. We have

$$-X_{F_i}|_W = \sum_{t=1}^{n} \sum_{m=1}^{l} \sum_{r,s=i_m+1}^{i_{m+1}} \{V_t, f_{rs}\}(p_0) V_r V_s \frac{\partial}{\partial V_t},$$

which is equal to

$$\sum_{m=1}^{l} \sum_{r,s,t=i_m+1}^{i_{m+1}} \{V_t, f_{rs}\}(p_0) V_r V_s \frac{\partial}{\partial V_t}$$

by virtue of Lemma 1.2.5. Hence the proposition follows. □

We now state a technical lemma which plays a key role in the study of the structure at a singular point. The proof will be given in Section 1.4.

LEMMA 1.2.7. *Let W be an n-dimensional vector space over $\mathbf{R}$, equipped with a positive definite inner product $<,>$. Let $\mathcal{G}$ be a vector space of W-valued quadratic forms on W satisfying the following conditions:*

1. *For any fixed $v \in W$, the quadratic forms $w \mapsto < G(w), v >$ $(G \in \mathcal{G})$ are simultaneously normalizable by an orthonormal basis of W;*
2. *elements of $\mathcal{G}$ are, regarded as vector fields on W, mutually commutative with respect to the Lie bracket;*
3. *$< G(w), w >= 0$ for any $w \in W$ and $G \in \mathcal{G}$.*

Then we get the following:

(a) *If $n = 2$, then $\dim \mathcal{G} \leq 1$ and $G \in \mathcal{G}$ is written as*

$$(ax_1 + bx_2)\left(x_2\frac{\partial}{\partial x_1} - x_1\frac{\partial}{\partial x_2}\right), \qquad a, b \in \mathbf{R},$$

where (x_1, x_2) is an orthonormal coordinate system on W;

(b) *if $n \geq 3$, then $\dim \mathcal{G} \leq n - 2$.*

Using Lemma 1.2.7 we get the following

PROPOSITION 1.2.8. *k $(= n - \dim \mathcal{F}_{p_0}) \leq n/2$, $l = n - k$, and changing the order of $W_1, \ldots, W_l$ suitably, we have*

$$\dim W_r = \begin{cases} 2 & \text{if } 1 \leq r \leq k \\ 1 & \text{if } k+1 \leq r \leq n-k. \end{cases}$$

Moreover, changing the basis $F_1, \ldots, F_n$ of $\mathcal{F}$ and the orthonormal basis V_{2r-1}, V_{2r} of W_r^* $(1 \leq r \leq k)$ suitably, we have

$$(F_r)_{p_0} = 0, \quad (F_{r+k})_{p_0} = V_{2r-1}^2 + V_{2r}^2 \qquad (1 \leq r \leq k),$$

$$X_{F_r}|_W = V_{2r}\left(V_{2r}\frac{\partial}{\partial V_{2r-1}} - V_{2r-1}\frac{\partial}{\partial V_{2r}}\right) \qquad (1 \leq r \leq k),$$

$$(F_r)_{p_0} = V_r^2 \qquad (2k+1 \leq r \leq n).$$

PROOF. We have already seen that $X_{F_i}|_W$ belongs to the subspace

$$\sum_{r=1}^{l} \mathcal{S}^2 W_r^* \otimes W_r$$

of $\mathcal{S}^2 W^* \otimes W$ for any i $(1 \leq i \leq k)$. Let $P_r(X_{F_i}|_W)$ denote the $\mathcal{S}^2 W_r^* \otimes W_r$-component of $X_{F_i}|_W$ and let $\mathcal{Q}_r$ be the subspace of $\mathcal{S}^2 W_r^* \otimes W_r$ spanned by them. We want to apply Lemma 1.2.7 to W_r and $\mathcal{Q}_r$. Since $X_{F_i}E = 0$, the vector fields $P_r(X_{F_i}|_W)$ belong to the kernel of $d(\sum_{t=i_r+1}^{i_{r+1}} V_t^2)$. Hence if $\dim W_r = 1$, we have $P_r(X_{F_i}|_W) = 0$, and if $\dim W_r \geq 2$, then the condition (3) in Lemma 1.2.7 is satisfied.

Now let us verify the conditions (1) and (2) in Lemma 1.2.7 in case $\dim W_r \geq 2$. Since $[X_{F_i}, X_{F_j}] = 0$, we have

$$[X_{F_i}|_W, X_{F_j}|_W] = 0 \qquad (1 \leq i, j \leq k).$$

But clearly

$$[P_r(X_{F_i}|_W), P_s(X_{F_j}|_W)] = 0$$

if $r \neq s$. Hence

$$[P_r(X_{F_i}|_W), P_r(X_{F_j}|_W)] = 0,$$

and the condition (2) is satisfied. To verify the condition (1), take a function

$$Y = \sum_{i=i_r+1}^{i_{r+1}} c_i V_i, \qquad c_i \in \boldsymbol{R}$$

around $W = T^*_{p_0} M$. Describing F_i as

$$F_i = \sum_{r=1}^{l} \sum_{s,t=i_r+1}^{i_{r+1}} f^i_{st} V_s V_t,$$

we get

$$P_r(X_{F_i}|_W) = - \sum_{m,s,t=i_r+1}^{i_{r+1}} \{V_m, f^i_{st}\} V_s V_t \frac{\partial}{\partial V_m}$$

and

$$\{Y, F_i\}_{p_0} = - \left\langle P_r(X_{F_i}|_W), \sum_{m=i_r+1}^{i_{r+1}} c_m \frac{\partial}{\partial V_m} \right\rangle,$$

where $<,>$ stands for the metric on W.

Now put $A_i = (f^i_{st})_{i_r+1 \le s,t \le i_{r+1}}$. Then A_i is a matrix-valued function around p_0, and the commutator $[A_i, A_j]$ vanishes. Regarding $Y = \sum c_m V_m$ as a vector field around p_0, we differentiate $[A_i, A_j]$ twice by Y at p_0. Then we get

$$(1.2.2) \qquad [YA_i, YA_j] = 0 \qquad \text{at } p_0.$$

Since $(YA_i)(p_0)$ is nothing but the matrix expression of the endomorphism

$$\{Y, F_i\}^e_{p_0},$$

it follows that the condition (1) is also satisfied.

Thus we have $Q_r = \{0\}$ if $\dim W_r = 1$, and by Lemma 1.2.7, $\dim Q_r \le \dim W_r - 2$ if $\dim W_r \ge 3$ and $\dim Q_r \le 1$ if $\dim W_r = 2$. Since

$$X_{F_i}|_W \in Q_1 \oplus \cdots \oplus Q_l \qquad (1 \le i \le k),$$

it follows that

$$k \le \sum_{r=1}^{l} \dim Q_r \le \sum_{r=1}^{l} (\dim W_r - 1) = n - l \le k.$$

The last inequality is due to Lemma 1.2.2. Thus the equality holds at each step in the inequalities above. Consequently we get $n - k = l$,

$$\sum_{i=1}^{k} \boldsymbol{R} X_{F_i}|_W = \sum_{r=1}^{n-k} Q_r,$$

$\dim Q_r = \dim W_r - 1$, $\dim W_r \le 2$, and the number of r such that $\dim W_r = 2$ is equal to k. Especially we have $2k \le n$. Changing the order, we may assume that

$$\dim W_r = 2 \quad \text{if} \quad 1 \le r \le k, \qquad = 1 \quad \text{if} \quad k+1 \le r \le n-k.$$

Moreover replacing $F_1, \ldots, F_k$ with their linear combinations we may assume that

$$X_{F_r}|_W = P_r(X_{F_r}|_W) \in Q_r \qquad (1 \le r \le k).$$

Then again by the previous lemma we can replace the orthonormal basis V_{2r-1}, V_{2r} of W_r^* appropriately and F_r by its constant multiple so that

$$X_{F_r}|_W = V_{2r}\left(V_{2r}\frac{\partial}{\partial V_{2r-1}} - V_{2r-1}\frac{\partial}{\partial V_{2r}}\right) \qquad (1 \le r \le k).$$

Finally, the functions $(F_{k+1})_{p_0}, \ldots, (F_n)_{p_0}$ now being linear combinations of

$$V_{2r-1}^2 + V_{2r}^2 \quad (1 \le r \le k), \qquad V_r^2 \quad (2k+1 \le r \le n),$$

we can replace $F_{k+1}, \ldots, F_n$ with their linear combinations so that

$$(F_{k+r})_{p_0} = V_{2r-1}^2 + V_{2r}^2 \quad (1 \le r \le k), \qquad (F_r)_{p_0} = V_r^2 \quad (2k+1 \le r \le n).$$

□

1.3. Local structure around a point in M^s

Let F_i, W_i, V_r be just as in Proposition 1.2.8. As before, W_i are extended to subbundles $\overline{W}_i$, and V_r are extended to an orthonormal frame on a neighborhood of p_0. In this case $\overline{W}_r^*$ are spanned by V_{2r-1} and V_{2r} if $1 \le r \le k$, and by V_{r+k} if $k+1 \le r \le n-k$ at each point on U. Take a small neighborhood U of p_0, and let I_r $(1 \le r \le k)$ be the maximal integral manifold in U of the integrable differential system (subbundle)

$$\sum_{\substack{1 \le s \le k \\ s \ne r}} \overline{W}_s^* + \sum_{s=k+1}^{n-k} \overline{W}_s^*$$

passing through p_0. Note that $\dim I_r = n-2$.

PROPOSITION 1.3.1. *If the neighborhood U of p_0 is sufficiently small, then*

$$I_r = \{p \in U \mid (F_r)_p = 0\} \qquad (1 \le r \le k).$$

PROOF. It is enough to show for I_1. We denote by F_r^s (resp. E^s) the $\mathcal{S}^2\overline{W}_s^*$-component of F_r (resp. E). F_r^s is written in the form

$$F_r^s = \begin{cases} h_r^{2s-1,2s-1}V_{2s-1}^2 + 2h_r^{2s-1,2s}V_{2s-1}V_{2s} + h_r^{2s,2s}V_{2s}^2 & (1 \le s \le k), \\ h_r^{k+s,k+s}V_{k+s}^2 \quad (k+1 \le s \le n-k), \end{cases}$$

where $h_r^{\alpha\beta} \in C^\infty(U)$. Let L_r $(2 \le r \le k+1)$ denote the maximal integral manifold in U of the differential system

$$\overline{W}_r^* + \cdots + \overline{W}_{n-k}^*$$

passing through p_0. Clearly we have

$$L_{k+1} \subset L_k \subset \cdots \subset L_2 = I_1.$$

We shall inductively show the following:

$(*)_r$ If $p \in L_r$, then $(F_t^s)_p = 0$ for any $s \ge r-1$ and $t \le r-1$.

Note that $(*)_2$ implies that I_1 is contained in the set

$$\{p \in U \mid (F_1)_p = 0\}.$$

First we shall show $(*)_{k+1}$. Fix $t \le k$. Since $\{E, F_t\} = 0$, it follows that

$$\{E^r, F_t^s\} + \{E^s, F_t^r\} = 0, \qquad (k+1 \le r, s \le n-k).$$

This implies that $\{V_{k+r}, h_t^{k+s,k+s}\}$ is a linear combination of $h_t^{2k+1,2k+1}, \cdots, h_t^{n,n}$ for any $r, s \ge k+1$. Since $h_t^{k+s,k+s}$ $(s \ge k+1)$ vanish at p_0, and since $V_{2k+1}, \ldots, V_n$ is a frame of L_{k+1}, it therefore follows that

$$h_t^{2k+1,2k+1} = \cdots = h_t^{n,n} = 0 \qquad \text{on } L_{k+1}.$$

In order to show $(*)_{k+1}$, we need to prove $(F_t^k)_p = 0$, $p \in L_{k+1}$. Restricting the formula

$$\{E^k, F_t^s\} + \{E^s, F_t^k\} = 0 \qquad (s \ge k+1)$$

to L_{k+1}, and observing the linear terms in the variable V_s in this formula, we see that

$$\{V_s, h_t^{\alpha\beta}\} \qquad (\alpha, \beta = 2k-1, 2k)$$

are written as linear combinations of $h_t^{\gamma,\delta}$ $(\gamma, \delta = 2k-1, 2k)$ on L_{k+1}. Since $(F_t^k)_{p_0} = 0$, it follows that $(F_t^k)_p = 0$ for $p \in L_{k+1}$.

Now fix $r \le k$ and assume that $(*)_{s+1}$ is true if $r \le s \le k$. We shall show $(*)_r$. For each $p \in L_{r+1}$ let $B_r(p)$ denote the maximal integral manifold in U of the differential system $\overline{W}_r^*$ passing through p. Taking U sufficiently small, we may assume that

$$B_r(p) \cap L_{r+1} = \{p\}$$

for any $p \in L_{r+1}$.

LEMMA 1.3.2. *If U is sufficiently small, then $(F_r^r)_q$ and $(E^r)_q$ are linearly independent for any $q \in B_r(p) - \{p\}$, $p \in L_{r+1}$.*

PROOF OF LEMMA 1.3.2. Since $(F_r^r)_p = 0$ for $p \in L_{r+1}$, it follows that $X_{F_r^r}$ is tangent to the fibre T_p^*M at each point on T_p^*M. Hence, choosing an orthonormal frame V_{2r-1}, V_{2r} of $\overline{W}_r^*$ suitably, we may assume that

$$\begin{aligned} &\text{the } \left(\frac{\partial}{\partial V_{2r-1}}, \frac{\partial}{\partial V_{2r}}\right)\text{-component of } X_{F_r^r}|_{T_r^* M} \\ &= c(p) V_{2r}\left(V_{2r}\frac{\partial}{\partial V_{2r-1}} - V_{2r-1}\frac{\partial}{\partial V_{2r}}\right), \end{aligned}$$

where $p \in L_{r+1}$ and $c \in C^\infty(L_{r+1})$. Since $c(p_0) \ne 0$, we may assume that $c(p) \ne 0$ for every $p \in L_{r+1}$. This implies that at $p \in L_{r+1}$,

$$\{V_{2r-1}, h_r^{2r-1,2r-1}\} = 0, \qquad \{V_{2r-1}, h_r^{2r-1,2r}\} = 0, \qquad \{V_{2r-1}, h_r^{2r,2r}\} = -c,$$

$$\{V_{2r}, h_r^{2r-1,2r-1}\} = 0, \qquad \{V_{2r}, h_r^{2r-1,2r}\} = \frac{c}{2}, \qquad \{V_{2r}, h_r^{2r,2r}\} = 0.$$

Therefore for a curve $u(t)$ in $B_r(p)$ such that

$$u(0) = p, \quad \dot{u}(0) = aV_{2r-1} + bV_{2r}, \quad a^2 + b^2 = 1,$$

we get

$$\frac{d^2}{dt^2}((h_r^{2r-1,2r-1} - h_r^{2r,2r})^2 + (h_r^{2r-1,2r})^2)(u(t))\,|_{t=0} = 2c^2 > 0.$$

This implies that

$$((h_r^{2r-1,2r-1} - h_r^{2r,2r})^2 + (h_r^{2r-1,2r})^2)(q) \geq \frac{c^2}{2} d(p,q)^2$$

for any $q \in B_r(p)$, provided U sufficiently small. Here d denotes the distance function of M. Since the left term of the inequality above is equal to the square of the difference of the two eigenvalues of $(F_r^r)_q^e$, the lemma follows. □

Since $(F_r^r)_q^e$ and $(F_t^r)_q^e$ are mutually commutative for $q \in B_r(p)$, this lemma implies that $(F_t^r)_q$ is a linear combination of $(E^r)_q$ and $(F_r^r)_q$ for any $t \leq r-1$ and $q \in B_r(p) - \{p\}$, $p \in L_{r+1}$;

$$(F_t^r)_q = a(q)(V_{2r-1}^2 + V_{2r}^2)_q + b(q)(F_r^r)_q.$$

LEMMA 1.3.3. *$a = 0$, and b is constant along $B_r(p)$.*

PROOF OF LEMMA 1.3.3. The constancy of a and b is an easy consequence of the fact that F_t^r, F_r^r, and E^r are mutually commutative with respect to the Poisson bracket. Since

$$(F_t^r)_p = (F_r^r)_p = 0,$$

we get $a = 0$ by taking a limit $q \to p$. □

This lemma shows that for each $t \leq r-1$,

$$F_t^r = bF_r^r \qquad \text{on} \quad L_r = \cup_{p \in L_{r+1}} B_r(p),$$

and $b \in C^\infty(L_r)$ which is constant on each fibre $B_r(p)$. Now we shall show that $b = 0$. For this purpose it suffices to prove that

$$\{F_t^r, V_{2r-1}\}_p = \{F_t^r, V_{2r}\}_p = 0 \quad \text{for any } p \in L_{r+1},$$

or equivalently,

$$\{V_u, h_t^{\alpha\beta}\} = 0 \quad \text{on } L_{r+1} \qquad (u, \alpha, \beta = 2r-1, 2r).$$

Since $\{E^r, F_t^s\} + \{E^s, F_t^r\} = 0$ $(s \geq r+1)$, it follows that on U, the functions

$$\{V_u, h_t^{\gamma\delta}\} \qquad (u = 2r-1, 2r,\ \gamma, \delta \geq 2r+1),$$

$$\{V_s, h_t^{\alpha\beta}\} \qquad (s \geq 2r+1,\ \alpha, \beta = 2r-1, 2r)$$

are linear combinations of $h_t^{\gamma\delta}$ $(\gamma, \delta \geq 2r-1)$ with coefficients in $C^\infty(U)$. Especially they vanish on L_{r+1}. Then by restricting the identity

$$\{V_s\{V_u, h_t^{\alpha\beta}\}\} = \{\{V_s, V_u\}, h_t^{\alpha\beta}\} + \{V_u, \{V_s, h_t^{\alpha\beta}\}\} \quad (s \geq 2r+1,\ u, \alpha, \beta = 2r-1, 2r)$$

to L_{r+1}, it turns out that $\{V_s, \{V_u, h_t^{\alpha\beta}\}\}$ are linear combinations of

$$\{V_{u'}, h_t^{\alpha'\beta'}\} \qquad (u', \alpha', \beta' = 2r-1, 2r)$$

on L_{r+1}. Also, since $X_{F_t} V_{2r-1} = X_{F_t} V_{2r} = 0$ at p_0, we have

$$\{V_u, h_t^{\alpha\beta}\} = 0 \qquad (u, \alpha, \beta = 2r-1, 2r)$$

at p_0. Therefore it follows that $\{V_u, h_t^{\alpha\beta}\} = 0$ on L_{r+1}, and consequently

$$(F_t^r)_q = 0 \quad \text{for any } q \in L_r, \quad t \leq r-1.$$

Now we know that

$$\{V_u, h_t^{\gamma\delta}\} \qquad (u = 2r-1, 2r, \quad \gamma, \delta \geq 2r+1)$$

are linear combinations of $h_t^{\gamma'\delta'}$ $(\gamma', \delta' \geq 2r+1)$, which vanish on L_{r+1}. Hence we have

$$(F_t^s)_q = 0 \quad \text{for any } q \in L_r, \ s \geq r+1, \ t \leq r-1.$$

Finally we get $(F_t^{r-1})_q = 0$ for all $q \in L_r$ and $t \leq r-1$, because

$$\{E^{r-1}, F_t^s\} + \{E^s, F_t^{r-1}\} = 0 \qquad (s \geq r),$$

and $(F_t^{r-1})_{p_0} = 0$. Hence we have shown $(*)_r$, and the induction has been completed.

In order to complete the proof of Proposition 1.3.1 it suffices to show that $(F_1)_q \neq 0$ if $q \in U - I_1$. It is shown, however, just in the same way as Lemma 1.3.2. □

Let us define the norm $|F_i^r|_q$ $(1 \leq r \leq n-k, \ 1 \leq i \leq n, \ q \in U)$ by

$$|F_i^r|_q = \begin{cases} \sqrt{h_i^{2r-1,2r-1}(q)^2 + h_i^{2r,2r}(q)^2 + 2h_i^{2r-1,2r}(q)^2} & (1 \leq r \leq k) \\ |h_i^{k+r,k+r}(q)| & (k+1 \leq r \leq n-k). \end{cases}$$

We also put

$$|F_i|_q = \sqrt{\sum_{r=1}^{n-k} |F_i^r|_q^2}.$$

The neighborhood U of p_0 is regarded as a fibre bundle over I_r $(1 \leq r \leq k)$. The fibres $B_r(p)$ $(p \in I_r)$ are the maximal integral manifolds of $\overline{W}_r^*$. We then define the function $|\cdot|_r$ on U by

$$|q|_r = d_{r,p}(q, p) \qquad \text{if} \quad q \in B_r(p),$$

where $d_{r,p}$ is the distance function on $B_r(p)$ with respect to the induced metric. We also put

$$|q|_0 = d(q, p_0), \qquad q \in U.$$

LEMMA 1.3.4. *If the neighborhood U of p_0 is sufficiently small, then there are constants $c, c_1 > 0$ such that the following inequalities hold on U.*

(1) $$(\operatorname{tr}(F_r^r)_q^e)^2 - 4\det(F_r^r)_q^e \geq c_1^2|q|_r^2, \quad |F_r^r|_q^2 \geq \frac{c_1^2}{2}|q|_r^2 \quad (1 \leq r \leq k).$$

(2) $$|F_{r+k}^r|_q \geq \frac{1}{2} \qquad (1 \leq r \leq n-k).$$

(3) $$|F_r^s|_q \leq c|q|_r^2 \qquad (1 \leq r \leq k, \ 1 \leq s \leq n-k, \ s \neq r),$$
$$|F_r^r|_q \leq c|q|_r \qquad (1 \leq r \leq k).$$

(4) $$|F_{r+k}^s|_q \leq c|q|_0 \qquad (1 \leq r, s \leq n-k, \ s \neq r).$$

PROOF. (1) Since

$$(\operatorname{tr}(F_r^r)^e)^2 - 4\det(F_r^r)^e = (h_r^{2r-1,2r-1} - h_r^{2r,2r})^2 + 4(h_r^{2r-1,2r})^2,$$

the first inequality has been already shown in the proof of Lemma 1.3.2. The second one follows from the inequality

$$|F_r^r|^2 \geq \frac{1}{2}(((F_r^r)^e)^2 - 4\det(F_r^r)^e).$$

(2) Since

$$(F_{r+k}^r)_{p_0} = V_{2r-1}^2 + V_{2r}^2 \quad (1 \leq r \leq k), \qquad = V_{r+k}^2 \quad (k+1 \leq r \leq n-k),$$

we have $|F_{r+k}^r|_{p_0} \geq 1$. Hence the inequality holds if q is close to p_0.

(3) Observe the linear terms in the variables V_{2r-1}, V_{2r} in the formula

$$\{E^r, F_r^s\} + \{E^s, F_r^r\} = 0.$$

Since $(F_r^r)_p = 0$ for $p \in I_r$, it follows that the first order derivatives of the coefficients of F_r^s by V_{2r-1}, V_{2r} also vanish at $p \in I_r$. This implies the first inequality. The second one follows from the fact that $(F_r^r)_p = 0$, $p \in I_r$.

(4) This inequality follows from the fact that $(F_{r+k}^s)_{p_0} = 0$. □

PROPOSITION 1.3.5. *Let $F \in \mathcal{F} - \{0\}$ be written as*

$$F = \sum_{i=1}^n a_i F_i, \qquad a_i \in \boldsymbol{R}.$$

Then on U we have:

(1) *If $a_i \neq 0$ for some $i \geq k+1$, then $F_q \neq 0$ for any $q \in U$;*
(2) *If $a_{k+1} = \cdots = a_n = 0$, $F_q = 0$, and $a_r \neq 0$, then $q \in I_r$ $(1 \leq r \leq k)$;*
(3) *$M^s \cap U = \cup_{r=1}^k I_r$.*

PROOF. Suppose that

$$F_q = \sum_{i=1}^n a_i (F_i)_q = 0$$

at a point $q \in U$. We shall apply Lemma 1.3.4 to this formula in several ways. First we have

$$\text{(1.3.1)} \qquad \frac{|a_{i+k}|}{2} \leq \sum_{1\leq r\leq k} c|a_r||q|_r^2 + \sum_{\substack{k+1\leq j\leq n\\ j\neq i+k}} c|a_j||q|_0 \quad (k+1 \leq i \leq n-k),$$

because $-a_{i+k}(F_{i+k}^i)_q = \sum_{j\neq i+k} a_j (F_j^i)_q$. Similarly,

$$\text{(1.3.2)} \qquad \frac{|a_{i+k}|}{2} \leq c|a_i||q|_i + \sum_{\substack{1\leq r\leq k\\ r\neq i}} c|a_r||q|_r^2 + \sum_{\substack{k+1\leq j\leq n\\ j\neq i+k}} c|a_j||q|_0 \quad (1 \leq i \leq k).$$

For $1 \leq i \leq k$ we take the difference of the eigenvalues of

$$a_i (F_i^i)_q^e = -\sum_{j\neq i} a_j (F_j^i)_q^e.$$

Since the summands are simultaneously diagonalizable, we have

$$|a_i|\sqrt{(\operatorname{tr}(F_i^i)_q^e)^2 - 4\det(F_i^i)_q^e} \le \sum_{j\neq i} |a_j|\sqrt{(\operatorname{tr}(F_j^i)_q^e)^2 - 4\det(F_j^i)_q^e}$$
$$\le \sqrt{2}\sum_{j\neq i} |a_j||F_j^i|_q.$$

This implies that

$$c_1|a_i||q|_i \le \sqrt{2}\sum_{\substack{1\le r\le k\\ r\neq i}} c|a_r||q|_r^2 + \sqrt{2}\sum_{k+1\le j\le n} c|a_j||q|_0 \quad (1\le i\le k). \tag{1.3.3}$$

We now put

$$\|q\|_0 = |q|_0 + \sum_{r+1}^{k} |q|_r, \qquad q\in U.$$

Then from (1.3.1) and (1.3.2),

$$\frac{|a_{i+k}|}{2} \le c\sum_{r=1}^{k} |a_r||q|_r + c\,\|q\|_0 \sum_{j=k+1}^{n} |a_j| \qquad (1\le i\le n-k). \tag{1.3.4}$$

Summing up both sides of this inequality with respect to i, we get

$$(\frac{1}{2} - (n-k)c\,\|q\|_0)\sum_{j=k+1}^{n} |a_j| \le (n-k)c\sum_{r=1}^{k} |a_r||q|_r. \tag{1.3.5}$$

From (1.3.3) we also get

$$(c_1 - \sqrt{2}kc\,\|q\|_0)\sum_{r=1}^{k} |a_r||q|_r \le \sqrt{2}kc\,\|q\|_0 \sum_{j=k+1}^{n} |a_j|. \tag{1.3.6}$$

Hence if U is sufficiently small so that the inequality

$$(\frac{1}{2} - (n-k)c\,\|q\|_0)(c_1 - \sqrt{2}kc\,\|q\|_0) > \sqrt{2}k(n-k)c^2\,\|q\|_0$$

holds for every $q\in U$, then we get

$$a_j = 0 \quad (k+1\le j\le n), \qquad |a_r||q|_r = 0 \quad (1\le r\le k).$$

Hence the proposition follows. □

The next corollary immediately follows from Proposition 1.3.5.

COROLLARY 1.3.6. *For $F\in\mathcal{F}-\{0\}$ the set $\{p\in M \mid F_p = 0\}$ is a disjoint union of closed submanifolds of even codimension, unless it is empty. Also, M^s is identical with the union of $(n-2)$-dimensional closed, connected submanifolds, each of which is a connected component of the set $\{p\in M\mid F_p = 0\}$ for some $F\in\mathcal{F}-\{0\}$. Moreover, if two of such $(n-2)$-dimensional submanifolds intersect, then at each intersection point their normal spaces are mutually orthogonal.*

Finally, we shall generalize Proposition 1.1.3 of Liouville and Stäckel to points on M^s. We go back to the situation before Corollary 1.3.6. Let $D_r(\epsilon)$ be the open ϵ-disk in C $(1\le r\le k)$, and let $z_r = x_{2r-1} + \sqrt{-1}x_{2r}$ be the natural coordinate function on it. Also, let L_s $(k+1 \leq s \leq n-k)$ be the open interval $(-\epsilon, \epsilon)$ and x_{k+s} its coordinate function. We put

$$R = D_1(\epsilon) \times \cdots \times D_k(\epsilon) \times L_{k+1} \times \cdots \times L_{n-k}.$$

PROPOSITION 1.3.7. *Let $F_1, \ldots, F_n$ be a basis of $\mathcal{F}$. If the neighborhood U of $p_0 \in M^s$ and $\epsilon > 0$ are suitably chosen, then there are C^∞ mapping $\Phi : R \to U$ and n^2 functions $b_{ij}(x_i)$ $(1 \leq i, j \leq n)$ possessing the following properties:*

(1) *There is a coordinate system on U so that U is identified with*

$$D_1(\epsilon^2) \times \cdots \times D_k(\epsilon^2) \times L_{k+1} \times \cdots \times L_{n-k} = \{(w_1, \ldots, w_k, x_{2k+1}, \ldots, x_n)\},$$

$p_0 = (0)$, and I_r is given by $w_r = 0$ $(1 \leq r \leq k)$;

(2) *$\Phi : R \to U$ is given by*

$$\Phi(z_1, \ldots, z_k, x_{2k+1}, \ldots, x_n) = (z_1^2, \ldots, z_k^2, x_{2k+1}, \ldots, x_n);$$

(3) *$b_{ij}(-x_i) = b_{ij}(x_i)$ for any i $(1 \leq i \leq 2k)$ and j;*

(4) *if the Taylor expansion of $b_{2r-1,j}(x_{2r-1})$ $(1 \leq r \leq k)$ at $x_{2r-1} = 0$ is given by*

$$b_{2r-1,j}(x_{2r-1}) \sim \sum_l \beta_{r,j;l}\, x_{2r-1}^{2l},$$

then at $x_{2r} = 0$,

$$b_{2r,j}(x_{2r}) \sim \sum_l (-1)^{l+1} \beta_{r,j;l}\, x_{2r}^{2l}.$$

(5) *by putting*

$$\gamma_{2r-1,j} = \beta_{r,j;1}, \quad \gamma_{2r,j} = \beta_{r,j;0} \quad (1 \leq r \leq k),$$
$$\gamma_{s,j} = b_{s,j}(0) \quad (2k+1 \leq s \leq n),$$

the determinant of the matrix (γ_{ij}) does not vanish;

(6) *the functions b_{ij} and the symmetric 2-tensor fields $(\partial/\partial x_i)^2$ are projectable with respect to Φ, and define C^∞ functions and C^∞ tensor fields on $U - M^s$; they satisfy*

$$\sum_j b_{ij}(x_i) F_j = \left(\frac{\partial}{\partial x_i}\right)^2.$$

Conversely, let U be the product space described in (1), and let R, $\Phi : R \to U$ be as above. Let $b_{ij}(x_i)$ be functions satisfying the conditions (3), (4) and (5) above. Then the formula

$$\sum_j b_{ij}(x_i) F_j = \left(\frac{\partial}{\partial x_i}\right)^2$$

defines C^∞ symmetric 2-tenstor fields $F_1, \ldots, F_n$ on a neighborhood of 0 on U via the branched covering $\Phi : R \to U$. Let $\mathcal{F}$ be the vector space spanned by $F_1, \ldots, F_n$. Then $\mathcal{F}$ satisfies the conditions (L.2), (L.3), and (L.4) in the definition of Liouville manifold on a neighborhood of (0), and satisfy the properness condition there. Moreover, $\dim \mathcal{F}_{(0)} = k$.

PROOF. Let U be a small neighborhood of p_0. Since the decomposition of TU to the sum of subbundles $\overline{W}_i^*$ gives the product structure on U, we can choose U so that it is identified with the product

$$B_1 \times \dots B_k \times B_{k+1} \times \dots B_{n-k},$$

where the factor B_r corresponds to the integral manifolds of $\overline{W}_r^*$, and we shall identify it with $B_r(p_0)$. We introduce an orientation on each B_r and fix it.

Let $F_1, \dots, F_n$ be the basis of $\mathcal{F}$ given in Proposition 1.2.8, and let F_s^r be, as before, the $S^2\overline{W}_r^*$-component of F_s. For each r $(1 \le r \le k)$ we write $F_r^r = h_{2r-1}V_{2r-1}^2 + h_{2r}V_{2r}^2$, where $h_{2r-1} \ge h_{2r}$ and V_{2r-1}, V_{2r} is a positive orthonormal (local) frame of $\overline{W}_r^*$ (different from the one previously used). Then, the functions h_{2r-1} and h_{2r} are continuous on U, of C^∞ on $U - I_r$, and satisfy $h_{2r-1} > h_{2r}$ on $U - I_r$ by virtue of Lemma 1.3.2. Moreover, V_{2r-1}^2, V_{2r}^2, and $V_{2r-1}V_{2r}$ are well-defined and of C^∞ on $U - I_r$. Also, let V_{k+s} $(k+1 \le s \le n-k)$ be a C^∞ section of unit length of $\overline{W}_s^*$ on U. Let a_{ij} $(1 \le i, j \le n)$ be the functions on $U - M^s$ defined by the formula $V_i^2 = \sum a_{ij}F_j$. We put

$$\widetilde{E}^r = \frac{V_{2r-1}^2}{a_{2r-1,r}} - \frac{V_{2r}^2}{a_{2r,r}} \qquad (1 \le r \le k).$$

LEMMA 1.3.8.

(1) $(h_{2r-1} - h_{2r})^{-1}\widetilde{E}^r$ *is of* C^∞ *and positive definite on each* $B_r(p)$ $(p \in I_r)$. *Thus it defines a riemannian metric on* $B_r(p)$ *via the "Legendre transform".*

(2) *The conformal structure on the factor* B_r *defined by the metric on* $B_r(p)$ *given in* (1) *is independent of* $p \in I_r$. *Hence it, together with the orientation, uniquely determines the complex structure* I *on* B_r $(1 \le r \le k)$.

(3) *The vector field* $V_s/\sqrt{a_{s,s}}$ $(2k+1 \le s \le n)$ *is well-defined and of* C^∞ *on* U, *and is not zero at* p_0.

PROOF OF LEMMA 1.3.8. As observed in the proof of Proposition 1.3.1, F_s^r is written as a linear combination of F_r^r and E^r with C^∞ coefficients for any s and r $(1 \le r \le k)$. Hence we can write

$$F_i = \sum_{r=1}^{k}(c_{i,r}F_r^r + 2c_{i,k+r}E^r) + \sum_{s=k+1}^{n-k} 2c_{i,k+s}E^s \qquad (1 \le i \le n),$$

$c_{ij} \in C^\infty(U)$. Since $c_{ij}(p_0) = \delta_{ij}$ if $i \le j$, the inverse matrix $(d_{ij}) = (c_{ij})^{-1}$ exists and $d_{ii} > 0$ on U, provided U small enough. Then,

$$F_r^r = \sum_j d_{r,j}F_j = h_{2r-1}V_{2r-1}^2 + h_{2r}V_{2r}^2 \quad (1 \le r \le k),$$

$$2E^r = \sum_j d_{k+r,j}F_j = V_{2r-1}^2 + V_{2r}^2 \quad (1 \le r \le k),$$

$$2E^s = \sum_j d_{k+s,j}F_j = V_{k+s}^2 \quad (k+1 \le s \le n-k).$$

Hence we have

$$
\begin{aligned}
a_{2r-1,j} &= \frac{d_{r,j} - h_{2r}d_{k+r,j}}{h_{2r-1} - h_{2r}} \quad (1 \le r \le k), \\
a_{2r,j} &= \frac{-d_{r,j} + h_{2r-1}d_{k+r,j}}{h_{2r-1} - h_{2r}} \quad (1 \le r \le k), \\
a_{s,j} &= d_{s,j} \qquad (2k+1 \le s \le n).
\end{aligned} \tag{1.3.7}
$$

Thus we have the formula

$$\frac{\widetilde{E}^r}{h_{2r-1} - h_{2r}} = \frac{2d_{r,r}E^r - d_{k+r,r}F^r_r}{(d_{r,r} - h_{2r}d_{k+r,r})(d_{r,r} - h_{2r-1}d_{k+r,r})}.$$

Since $h_{2r-1} + h_{2r}$ and $h_{2r-1}h_{2r}$ are C^∞ functions on U, (1) follows.

To prove (2) we note that the symmetric tensor fields $V_{2r-1}^2/a_{2r-1,r}$, $V_{2r}^2/a_{2r,r}$ $(1 \le r \le k)$, and $V_s^2/a_{s,s}$ $(2k+1 \le s \le n)$ on $U - M^s$ are mutually commutative with respect to the Poisson bracket (cf. Proposition 1.1.3). Therefore, the riemannian metric on $B_r - \{p_0\}$ defined by $\widetilde{E}^r$ on $B_r(p) - \{p\}$ $(p \in I_r)$ does not depend on p. Thus (2) follows. The assertion (3) is obvious, because $a_{s,s}$ $(2k+1 \le s \le n)$ are positive C^∞ functions on U. □

Let w_r $(1 \le r \le k)$ be a complex coordinate function on B_r satisfying $w_r(p_0) = 0$, and let $D_r(\epsilon) \to B_r$ be the holomorphic mapping given by $w_r = z_r^2$. Put

$$X_r = \frac{V_{2r-1}}{\sqrt{a_{2r-1,r}}} - \sqrt{-1}\frac{V_{2r}}{\sqrt{-a_{2r,r}}}.$$

Since X_r is well-defined up to sign on $B_r - \{p_0\}$, it lifts to $D_r(\epsilon) - \{0\}$ as a single-valued vector field. Note that it is holomorphic, because

$$\left[\frac{V_{2r-1}}{\sqrt{a_{2r-1,r}}}, \frac{V_{2r}}{\sqrt{-a_{2r,r}}}\right] = 0.$$

We claim that the holomorphic vector field X_r extends to $0 \in D_r(\epsilon)$ and is not zero at 0. In fact, by the proof of Lemma 1.3.2 we have

$$c|w_r| < h_{2r-1} - h_{2r} < c'|w_r|$$

for some positive constant c, c'. This implies that $\widetilde{E}^r = X_r\overline{X}_r$, lifted to $D_r(\epsilon) - \{0\}$, satisfies the inequalities

$$c_1\frac{\partial}{\partial z_r}\frac{\partial}{\partial \overline{z}_r} < \widetilde{E}^r < c_1'\frac{\partial}{\partial z_r}\frac{\partial}{\partial \overline{z}_r}$$

around $0 \in D_r(\epsilon)$ for some positive constants c_1 and c_1'. Thus X_r extends to $0 \in D_r(\epsilon)$ by elementary function theory, and it does not vanish at 0.

Therefore, there is a complex coordinate function z_r' on a neighborhood of $0 \in D_r(\epsilon)$ such that

$$z_r'(0) = 0, \qquad X_r = 2\frac{\partial}{\partial z_r'}.$$

Rewriting z_r' as z_r and $w_r' = {z_r'}^2$ as w_r, we have another coordinate function w_r on B_r. Also, we define a coordinate function x_s on B_{s-k} $(2k+1 \le s \le n)$ by

$$x_s(p_0) = 0, \qquad \frac{V_s}{\sqrt{a_{s,s}}} = \frac{\partial}{\partial x_s}.$$

Then we have a mapping $\Phi : R \to U$ defined by (2), and redefining U as the image of Φ, we have the identification in (1).

We put

$$b_{2r-1,j} = \frac{a_{2r-1,j}}{a_{2r-1,r}}, \quad b_{2r,j} = -\frac{a_{2r,j}}{a_{2r,r}} \quad (1 \le r \le k), \quad b_{s,j} = \frac{a_{s,j}}{a_{s,s}} \quad (2k+1 \le s \le n).$$

Then we have

$$\sum_j b_{i,j} F_j = \left(\frac{\partial}{\partial x_i}\right)^2$$

on $U - M^s$. By virtue of Proposition 1.1.3, we see that the function $b_{i,j}$, lifted to R, depends only on the variable x_i, which is a C^∞ function on L_{i-k} if $2k+1 \le i \le n$, and is an even C^∞ function where $x_i \ne 0$ if $1 \le i \le 2k$. By the formula (1.3.7), to prove that $b_{i,j}(x_i)$ is also of C^∞ at $x_i = 0$ it suffices to show that h_{2r-1} and h_{2r} are, when lifted, C^∞ functions on R.

LEMMA 1.3.9. *The function $(h_{2r-1} - h_{2r})/|z_r|^2$ is a positive C^∞ function on U.*

PROOF OF LEMMA 1.3.9. Since $\widetilde{E}^r$ is C^∞ on R and positive definite on the factor $D_r(\epsilon)$, it follows that the Φ_*-image of $\widetilde{E}^r/|z_r|^2$ is C^∞ on U and positive definite on the factor B_r. Hence, by Lemma 1.3.8 (1) we see that the function $(h_{2r-1} - h_{2r})/|z_r|^2$ is C^∞ and positive on U. □

Since the sum $h_{2r-1} + h_{2r}$ is C^∞, it thus follows that h_{2r-1} and h_{2r} are C^∞ functions on R. By (1.3.7) we have

$$\frac{b_{2r-1,j} + b_{2r,j}}{|z_r|^2} = \frac{h_{2r-1} - h_{2r}}{|z_r|^2} \frac{d_{r,r}d_{k+r,j} - d_{r,j}d_{k+r,r}}{(d_{r,r} - h_{2r}d_{k+r,r})(d_{r,r} - h_{2r-1}d_{k+r,r})}. \tag{1.3.8}$$

Note that the right-hand side of the formula above is a C^∞ function on U. This implies that the Taylor expansion of the left-hand side at $0 \in D_r(\epsilon)$ is a formal power series of $x_{2r-1}^2 - x_{2r}^2$ and $x_{2r-1}x_{2r}$, which indicates (4). The assertion (5) easily follows from the formula (1.3.8) and the fact that $b_{2r-1,j} - b_{2r,j} = 2d_{r,j}$ at $0 \in D_r(\epsilon)$.

The converse is also proved by the observation above. □

1.4. Proof of Lemma 1.2.7

First we consider the case $n = 2$. Let $G \in \mathcal{G}$. Then by the condition (3) G should be of the form

$$G = (ax_1 + bx_2)\left(x_2\frac{\partial}{\partial x_1} - x_1\frac{\partial}{\partial x_2}\right),$$

where $a, b \in \boldsymbol{R}$ and (x_1, x_2) is an orthonormal coordinates on W. If G' is another element of $\mathcal{G}$, then it has a similar form. Thus by the condition (2) we see that G and G' are linearly dependent.

Now, let us assume that $n \ge 3$. Let $x_1, \dots, x_n$ be an orthonormal linear coordinate system on W. We assume that $\dim \mathcal{G} \ge n-1$, and derive a contradiction. Take a subspace if necessary, we may assume that $\dim \mathcal{G} = n-1$. We consider the following two cases separately:

1. $\dim\{G(v) \mid G \in \mathcal{G}\} = n-1$ for some $v \in W$;
2. $\dim\{G(v) \mid G \in \mathcal{G}\} \le n-2$ for any $v \in W$.

Case 1. We may assume that $\dim\{G(v_0) \mid G \in \mathcal{G}\} = n-1$, where $v_0 = (0, \dots, 0, 1)$. Take a basis $H_1, \dots, H_{n-1}$ of $\mathcal{G}$ such that $H_i(v_0) = \partial/\partial x_i$. Then, since $< H_i(x), x >= 0$ for any $x \in W$, it follows that H_i is of the form

$$H_i = x_n^2 \frac{\partial}{\partial x_i} - x_n x_i \frac{\partial}{\partial x_n} + x_n \sum_{k,l=1}^{n-1} \alpha^i_{kl} x_k \frac{\partial}{\partial x_l} - \sum_{k,l=1}^{n-1} \beta^i_{kl} x_k x_l \frac{\partial}{\partial x_n} + \sum_{j=1}^{n-1} h_{ij}(x) \frac{\partial}{\partial x_j},$$

where α^i_{kl}, $\beta^i_{kl} \in \boldsymbol{R}$, $\beta^i_{kl} = \beta^i_{lk}$, $h_{ij}(x)$ are quadratic forms not containing the variable x_n, and

$$\alpha^i_{kl} + \alpha^i_{lk} = 2\beta^i_{kl}, \qquad \sum_{j=1}^{n-1} h_{ij}(x) x_j = 0. \tag{1.4.1}$$

We shall show that

$$\alpha^i_{kl} = \beta^i_{kl} = 0$$

by using the conditions $[H_i, H_j] = 0$.

First, observing the terms containing x_n^3 in $[H_i, H_j]$, we get

$$\alpha^i_{jl} = \alpha^j_{il} \qquad \text{for any } i, j, l. \tag{1.4.2}$$

Next, observing the terms containing x_n^2 in $[H_i, H_j]$, we get

$$\alpha^j_{ki} - \alpha^i_{kj} - 2\beta^j_{ki} + 2\beta^i_{kj} = 0. \tag{1.4.3}$$

$$\frac{\partial}{\partial x_i} h_{jm} = \frac{\partial}{\partial x_j} h_{im} \qquad (m \neq i, j). \tag{1.4.4}$$

$$\frac{\partial}{\partial x_i} h_{ji} - \frac{\partial}{\partial x_j} h_{ii} + 2x_j = 0 \qquad (i \neq j). \tag{1.4.5}$$

By (1.4.1), (1.4.2), and (1.4.3) we have

$$\alpha^i_{jk} = \beta^i_{jk}, \quad \alpha^i_{jk} = \alpha^j_{ik} = \alpha^i_{kj} \tag{1.4.6}$$

Also, observing the terms containing $x_n \partial/\partial x_n$ in $[H_i, H_j]$, and using (1.4.6), we get

$$h_{ij} = h_{ji} \tag{1.4.7}$$

By (1.4.1), (1.4.4), (1.4.5), and (1.4.7) we have for $i \neq m$:

$$\begin{aligned} 0 &= \frac{\partial}{\partial x_m} \sum_{j=1}^{n-1} h_{ij} x_j \\ &= h_{im} + \sum_{j \neq m,i} (\frac{\partial}{\partial x_m} h_{ij}) x_j + (\frac{\partial}{\partial x_m} h_{im}) x_m + (\frac{\partial}{\partial x_m} h_{ii}) x_i \\ &= h_{im} + \sum_j (\frac{\partial}{\partial x_j} h_{im}) x_j - (\frac{\partial}{\partial x_i} h_{im}) x_i + (\frac{\partial}{\partial x_m} h_{ii}) x_i \\ &= 3h_{im} + 2x_m x_i. \end{aligned}$$

Hence

$$h_{ij} = -\frac{2}{3} x_i x_j \quad (i \neq j), \quad h_{ii} = \frac{2}{3} \sum_{m \neq i} x_m^2. \tag{1.4.8}$$

Now observe the terms containing $\partial/\partial x_n$ in $[H_i, H_j]$. Then one obtains

$$-\sum_{kl}\alpha^j_{kl}x_kx_lx_i + \sum_{kl}\alpha^i_{kl}x_kx_lx_j + 2\sum_{kl}\alpha^i_{kl}h_{jl}x_k - 2\sum_{kl}\alpha^j_{kl}h_{il}x_k = 0,$$

which, combined with (1.4.8), implies

$$\sum_{kl}\alpha^j_{kl}x_kx_lx_i = \sum_{kl}\alpha^i_{kl}x_kx_lx_j. \tag{1.4.9}$$

By (1.4.9) and (1.4.6) one can easily see that

$$\alpha^i_{jk} = 0 \quad \text{for any } i, j, k.$$

Hence we have

$$< H_i, \frac{\partial}{\partial x_n} >= -x_nx_i.$$

Since they are not simultaneously normalizable, we have a contradiction. This finishes the proof of Case 1.

Case 2. Suppose $\max_{v\in W} \dim\{H(v) \mid H \in \mathcal{G}\} = \alpha \ (\leq n-2)$. Choose an orthonormal coordinate system $(x_1, \ldots, x_n)$ of W and a basis $H_1, \ldots, H_{n-1}$ of $\mathcal{G}$ such that

$$H_i(v_0) = \frac{\partial}{\partial x_i} \quad (1 \leq i \leq \alpha), \qquad H_j(v_0) = 0 \quad (\alpha+1 \leq j \leq n-1),$$

$$v_0 = (0, \ldots, 0, 1) \in W.$$

Then describing

$$H_i = \sum_{j=1}^{n} H_{ij}(x)\frac{\partial}{\partial x_j},$$

we have

$$\begin{aligned} H_{ij}(x) &= \delta_{ij}x_n^2 + (\text{terms of lower degrees in } x_n) \\ H_{in}(x) &= -x_ix_n + (\text{terms not containing } x_n) \end{aligned}$$

for i, $1 \leq i \leq \alpha$. We shall show that $H_j = 0$ $(j \geq \alpha+1)$. We denote by $\overline{H}_{ij}$ the self-adjoint endomorphism of W determined by the formula:

$$< \overline{H}_{ij}v, v >= H_{ij}(v).$$

We first claim that

(1) $H_{jn}(x)$ $(\alpha+1 \leq j \leq n-1)$ does not contain the variables $x_1, \ldots, x_\alpha, x_n$.
In fact, since $H_j(v_0) = 0$, each H_{jk} does not contain x_n^2. Since $\sum_k H_{jk}(x)x_k = 0$, it turns out that H_{jn} does not contain the variable x_n. Hence we have

$$0 = \overline{H}_{in}\overline{H}_{jn}e_n = \overline{H}_{jn}\overline{H}_{in}e_n = -\frac{1}{2}\overline{H}_{jn}e_i,$$

where $e_1, \ldots, e_n$ is the basis of W dual to $x_1, \ldots, x_n$. Observing the formula $[H_i, H_j] = 0$ at v_0 we also have

(2) $H_{jk}(x)$ $(\alpha+1 \leq j \leq n-1$, any $k)$ does not contain x_nx_i $(1 \leq i \leq \alpha)$.
Next we claim that

(3) H_{jk} $(\alpha \leq j, k \leq n-1)$ does not contain the variable x_n.

In fact, assume that H_{jk} contains a term $x_n x_l$. By (2) we have $l \geq \alpha + 1$. Then clearly there is a point v_1 near v_0 such that $H_1(v_1), \dots, H_\alpha(v_1), H_j(v_1)$ are linearly independent, which contradicts the definition of α. We also have

(4) $H_{ji}(x)$ $(\alpha + 1 \leq j \leq n-1,\ 1 \leq i \leq \alpha)$ does not contain x_n.

In fact, assume that H_{ji} contains a term $x_n x_l$. By (2) we have $l \geq \alpha + 1$. Since $\sum_k H_{jk}(x) x_k = 0$, and since $H_{jn}(x)$ does not contain the variable x_i, it follows that $H_{jl}(x)$ contains the term $x_n x_i$. This, however, contradicts (3). By (3) and (4) we have just obtained

(5) $H_{jk}(x)$ $(\alpha + 1 \leq j \leq n-1$, any $k)$ does not contain the variable x_n.

Combined with the formula $\sum_k H_{jk}(x) x_k = 0$, this indicates

(6) $H_{jn} = 0$ $(\alpha + 1 \leq j \leq n-1)$.

By (6) we have

(7) $H_{ji}(x) = 0$ $(\alpha + 1 \leq j \leq n-1,\ 1 \leq i \leq \alpha)$.

In fact, observing the coefficient of $\partial/\partial x_n$ in the formula $0 = [H_j, H_i]$, we have $H_j H_{in} = 0$. Taking the terms containing x_n, we then have $H_{ji}(x) = 0$.

Finally we claim that

(8) $H_{jk}(x) = 0$ $(\alpha + 1 \leq j, k \leq n-1)$.

In fact, assume that it is not the case. Then, as is easily seen, there is a point $v_1 \in W$ near v_0 such that $H_1(v_1), \dots, H_\alpha(v_1), H_j(v_1)$ are linearly independent. This is a contradiction.

By (6), (7), and (8) we obtain

$$H_j = 0 \qquad (\alpha + 1 \leq j \leq n-1),$$

which contradicts $\dim \mathcal{G} = n - 1$. This completes the proof of Lemma 1.2.7.

2. Global Structure of Proper Liouville Manifolds

2.1. Submanifolds J

As stated in Introduction, we shall use the adjective *admissible* in the following sense: A subspace W of the cotangent space T_p^*M at a point $p \in M$ is called admissible if W is a direct sum of simultaneous eigenspaces of the endomorphisms F_p^e, $F \in \mathcal{F}$. Also, a subspace of the tangent space T_pM is called admissible if it is the dual space of an admissible subspace of T_p^*M, i.e., the orthogonal complement of the annihilator. An orthonormal basis of T_pM is called admissible if it normalizes every F_p, $F \in \mathcal{F}$. An orthonormal frame defined on an open subset of M^0 is called admissible if it is admissible at every point. We shall also say that an orthonormal frame on a neighborhood of $p \in M^s$ is *semi*-admissible if each of the n vector fields is a section of one of the subbundles $\overline{W}_j^*$ described after Lemma 1.2.3. Finally a submanifold N of M is called admissible if T_pN is an admissible subspace for every $p \in N$.

Let I be an $(n-2)$-dimensional, connected, closed submanifold of M contained in M^s. Then by Corollary 1.3.6 there is $F \in \mathcal{F}$ such that I is equal with one of the connected components of

$$\{p \in M \mid F_p = 0\}.$$

Put

$$J = \{p \in M \mid \operatorname{rank} F_p \leq 1\}.$$

The main purpose of this section is to prove the following

PROPOSITION 2.1.1. *J is a closed submanifold of codimension one that is connected and totally geodesic. Moreover, $F(\mu)=0$ and $(X_F)_\mu=0$ if $\mu\in T^*J$.*

Here T^*J is regarded as a submanifold of T^*M as stated in Preliminary remarks. The proof needs several lemmas.

LEMMA 2.1.2. *If $(X_F)_\lambda = 0$ at $\lambda\in S^*_{p_0}M$, $p_0\in I$, then the rank of the quadratic form $F_{\pi(\zeta_t\lambda)}$ is 0 or 1 for every $t\in \boldsymbol{R}$.*

PROOF. Assume first that $\dim \mathcal{F}_{p_0}=n-1$. (Note that the set of all such points $p_0\in I$ is an open and dense subset of I.) Let $V_1,\dots,V_n$ be an admissible orthonormal basis of $T_{p_0}M$ so that

$$X_F|_{T^*_{p_0}M}=cV_2\left(V_2\frac{\partial}{\partial V_1}-V_1\frac{\partial}{\partial V_2}\right),\quad c\in\boldsymbol{R}-\{0\}.$$

Also, let $F_1,\dots,F_n$ be a basis of $\mathcal{F}$ so that

$$F_1=F,\quad (F_2)_{p_0}=V_1^2+V_2^2,\quad (F_i)_{p_0}=V_i^2\quad (i\ge 3).$$

Now we assume that $\lambda\in S^*_{p_0}M$ satisfies

$$V_1(\lambda)\neq 0,\quad V_2(\lambda)=0,\quad V_i(\lambda)\neq 0\quad (i\ge 3).$$

Of course, the set of such λ is open and dense in

$$\{\mu\in S^*_{p_0}M\mid (X_F)_\mu=0\}=S^*_{p_0}M\cap V_2^{-1}(0).$$

Let us consider the Jacobi fields

$$Y_i(t)=\pi_*(X_{F_i})_{\zeta_t\lambda},\qquad 2\le i\le n,$$

along the geodesic $\gamma(t)=\pi(\zeta_t\lambda)$. Since

$$Y_2(0)=2V_1(\lambda)V_1,\quad Y_i(0)=2V_i(\lambda)V_i\quad (i\ge 3),$$

it follows that the vectors $Y_2(0),\dots,Y_n(0)$ are linearly independent. This implies, as the proof of Proposition 1.1.1, that the set of $t\in\boldsymbol{R}$ such that $Y_2(t),\dots,Y_n(t)$ are linearly dependent at $\gamma(t)$ is discrete. We put

$$W(t)=\sum_{i=2}^n \boldsymbol{R}Y_i(t)\quad\subset T_{\gamma(t)}M.$$

We shall show that $\operatorname{rank}F_{\gamma(t)}\le 1$ when $\dim W(t)=n-1$. First take $t_0\in\boldsymbol{R}$ such that $\dim W(t_0)=n-1$ and $\gamma(t_0)\in M^0$. Let $\overline{V}_1,\dots,\overline{V}_n$ be an admissible orthonormal frame on a neighborhood U of $q_0=\gamma(t_0)$. Then F_i are described as

$$(F_i)_q=\sum_{j=1}^n f_{ij}(q)\overline{V}_j^2,\qquad q\in U.$$

Since $\pi_*(X_{F_1})_{\zeta_{t_0}\lambda}=0$, we have

$$f_{1j}(q_0)\overline{V}_j(\zeta_{t_0}\lambda)=0$$

for any j. Changing the ordering of $\{\overline{V}_j\}$, we may assume that

$$\overline{V}_1(\zeta_{t_0}\lambda)=\cdots=\overline{V}_k(\zeta_{t_0}\lambda)=0,\quad \overline{V}_j(\zeta_{t_0}\lambda)\neq 0\quad (j\ge k+1).$$

Then $f_{1j}(q_0) = 0$ for $j \geq k+1$, and $\operatorname{rank}(F_1)_{q_0} \leq k$. On the other hand, since

$$Y_i(t_0) = \pi_*(X_{F_i})_{\zeta_{t_0}\lambda} = 2\sum_j f_{ij}(q_0)\overline{V}_j(\zeta_{t_0}\lambda)\overline{V}_j,$$

it follows that $n-1 = \dim W(t_0) \leq n-k$. Hence we have $k \leq 1$.

Next we consider the case where $\dim W(t_0) = n-1$ and $q_0 = \gamma(t_0) \in M^s$. In this case there is $G = \sum_i a_i F_i \in \mathcal{F} - \{0\}$ such that $G_{q_0} = 0$. Then we have

$$0 = \pi_*(X_G)_{\zeta_{t_0}\lambda} = \sum_{i\geq 2} a_i Y_i(t_0).$$

This implies that $a_2 = \cdots = a_n = 0$ and $G = a_1F_1$, $a_1 \neq 0$. Hence it follows that $(F_1)_{q_0} = 0$, and we consequently have $\operatorname{rank} F_{\gamma(t_0)} \leq 1$ under the assumption $\dim W(t_0) = n-1$.

Hence by continuity we have $\operatorname{rank}(F_1)_{\gamma(t)} \leq 1$ for every $t \in \boldsymbol{R}$. A continuity argument also implies that

$$\operatorname{rank}(F_1)_{\pi(\zeta_t\lambda)} \leq 1$$

for any $t \in \boldsymbol{R}$ and $\lambda \in S^*_{p_0}M$ such that $(X_{F_1})_\lambda = 0$. Finally, $p_0 \in I$ may also be arbitrary by the same reason. $\square$

Let N^*I denote the conormal bundle of I, and put

$$\widetilde{L}_p = \{\lambda \in N^*_pI \mid (X_F)_\lambda = 0\}, \qquad p \in I.$$

Then $\widetilde{L} = \cup_{p\in I}\widetilde{L}_p$ is a smooth line subbundle of N^*I. Let $p_0 \in I$. In view of Proposition 1.2.8 we can take a semi-admissible orthonormal frame $V_1, \ldots, V_n$ on a neighborhood of p_0 so that $V_3, \ldots, V_n$ form a frame of TI and

$$X_F|_{T^*_pM} = c(p)V_2\left(V_2\frac{\partial}{\partial V_1} - V_1\frac{\partial}{\partial V_2}\right), \qquad c(p) > 0,\ p \in I.$$

Here, note that V_1 is uniquely determined along I. Hence we obtain a globally defined vector field, say V_-, along I by putting $V_- = V_1$ at each point. This implies that the line bundle $\widetilde{L}$ is trivial, because it admits the everywhere non-zero section V_-^* that is dual to V_-.

Let ϵ be a positive continuous function on I such that the geodesic $t \mapsto \pi(\zeta_t\lambda)$ is minimal on the interval $|t| < \epsilon(p)$ for every $\lambda \in N^*_pI$, $2E(\lambda) = 1$, $p \in I$. We put

$$\begin{aligned} U &= \{\pi(\zeta_t\lambda) \mid \lambda \in N^*_pI,\ 2E(\lambda) = 1,\ |t| < \epsilon(p),\ p \in I\} \\ L_\pm &= \{\pi(\zeta_t\lambda) \mid \lambda \in \widetilde{L}_p,\ V_-(\lambda) = \mp 1,\ 0 < t < \epsilon(p),\ p \in I\} \\ L &= L_- \cup I \cup L_+. \end{aligned}$$

Then U is a neighborhood of I, and L is its submanifold of codimension one. Taking ϵ sufficiently small, we have the following

LEMMA 2.1.3. *For every $p \in L_+$ (resp. L_-), the quadratic form $(F)_p$ is of rank one and positive (resp. negative) semi-definite.*

PROOF. We have already seen in Lemma 2.1.2 that $\operatorname{rank} F_p \leq 1$ for $p \in L$. Fix $p_0 \in I$, and let $V_1, \ldots, V_n$ be a semi-admissible frame as above around p_0. We describe F as

$$F = h_{11}V_1^2 + h_{12}V_1V_2 + h_{22}V_2^2 + (\text{lower-degree-terms in } V_1,\ V_2)$$

on a neighborhood of p_0. Then computing $X_F|_{T^*_{p_0}M}$ we have

$$\{V_1, h_{11}\} = \{V_1, h_{12}\} = 0, \quad \{V_1, h_{22}\} = -c \quad \text{at } p_0.$$

This implies that for $\lambda \in \widetilde{L}_{p_0}$, $V_1(\lambda) = 1$, the function $h_{22}(\pi(\zeta_t\lambda))$ is negative on $0 < t < \epsilon(p_0)$, and positive on $-\epsilon(p_0) < t < 0$, provided $\epsilon(p_0)$ sufficiently small. This indicates the lemma. □

LEMMA 2.1.4. *The submanifold L is totally geodesic. Also, if $\mu \in T^*L$ ($\subset T^*M$), then $F(\mu) = (X_F)_\mu = 0$.*

PROOF. Fix p_0 and $q_0 = \pi(\zeta_{t_0}\lambda) \in L_-$, $0 < t_0 < \epsilon(p_0)$, where $\lambda \in \widetilde{L}_{p_0}$, $V_-(\lambda) = 1$. Take a point $p \in I$ sufficiently near p_0, and let $p_i \in L_+$ $(i = 1, 2, \ldots)$ be a sequence of points such that $p_i \to p$ as $i \to \infty$. We may assume that there are unique minimal geodesics $\gamma_i(t)$, $0 \leq t \leq t_i$, from q_0 to p_i and $\gamma_\infty(t)$, $0 \leq t \leq t_\infty$, from q_0 to p, and that $\gamma_i \to \gamma_\infty$ as $i \to \infty$.

We first claim that the vectors $\dot\gamma_i(0)$ and $\dot\gamma_i(t_i)$ are tangent to L. In fact, suppose that $\dot\gamma_i(t_i) \notin T_{p_i}L$. Since the exponential mapping Exp_{q_0} is a diffeomorphism around $\dot\gamma_i(0) \in T_{q_0}M$, it follows that the hypersurface $A \subset T_{q_0}M$ defined by $\mathrm{Exp}_{q_0}(A) \subset L$ intersects the line $\boldsymbol{R}\dot\gamma_i(0)$ transversally at $\dot\gamma_i(0)$. On the other hand, since $F_{q_0} \leq 0$ and $F_{p_i} \geq 0$, it follows that $F_{q_0}(A) = 0$. Hence we get $F_{q_0} = 0$, which contradicts Lemma 2.1.3. Therefore $\dot\gamma_i(t_i) \in T_{p_i}L$, and also $\dot\gamma_i(0) \in T_{q_0}L$ from the same argument.

Put $\mu = \sharp^{-1}(\dot\gamma_\infty(0)) \in S^*_{q_0}M$, where $\sharp : T^*M \to TM$ is the bundle isomorphism induced from the metric g. Since $\sharp(\zeta_{t_\infty}\mu) = \dot\gamma_\infty(t_\infty) \in T_pL$, we have $(X_F)_{\zeta_{t_\infty}\mu} = 0$, and hence

$$(X_F)_\mu = 0, \qquad F(\mu) = 0.$$

Since the mapping $I \ni p \mapsto \mu \in S^*_{q_0}L$ is a diffeomorphism around p_0, it thus follows that

$$(X_F)_\mu = 0, \quad F(\mu) = 0 \quad \text{for all } \mu \in T^*_{q_0}L.$$

Now assume that $q_0 \in M^0$. Since $F(\mu) = 0$, $\mu \in T^*_{q_0}L$, we can take an admissible orthonormal frame $V_1, \ldots, V_n$ around q_0 such that V_1 is orthogonal to L. Describing F as

$$F = \sum_{i=1}^{n} f_i V_i^2,$$

we get $f_2 = \cdots = f_n = 0$ on L and

$$0 = -(X_F)_\mu V_1 = \sum_i \{V_1, f_i\} V_i(\mu)^2$$

for any $\mu \in T^*_{q_0}L$. Hence $\{V_1, f_i\}(q_0) = 0$ for $2 \leq i \leq n$.

Let us now observe the formula $0 = \{F, E\}$. Observing the terms that are linear in V_1 and quadratic in V_i $(i \geq 2)$, we get

$$f_1 \times (\text{the } V_i\text{-component of } \{V_1, V_i\}) = 0 \quad \text{at } q_0.$$

Since $f_1(q_0) \neq 0$, this implies that

$$[V_1, V_i] \in \boldsymbol{R}V_1 \qquad \text{at } q_0$$

by regarding $V_1, \dots, V_n$ as vector fields. Hence the second fundamental form of L vanishes at q_0. Since $L \cap M^0$ is dense in L, it is also true for $q \in M^s$ by continuity.□

LEMMA 2.1.5. $\{q \in U \mid \operatorname{rank} F_q \leq 1\} = L$.

PROOF. Let $q_0 \in U - L$, and suppose that q_0 belongs to the fibre at $p_0 \in I$, i.e.,

$$q_0 = \pi(\zeta_{t_0}\lambda), \quad \lambda \in N^*_{p_0}I - \widetilde{L}_{p_0}, \quad 2E(\lambda) = 1, \quad |t_0| < \epsilon(p_0).$$

Take points $p_- \in L_-$ and $p_+ \in L_+$ sufficiently near p_0, and join them with the point q_0 by minimal geodesics γ_- and γ_+ respectively. Then the image of $\gamma_\pm$ is contained in U. Note that the geodesic $\gamma_\pm$ is not tangent to L at $p_\pm$. In fact if it is tangent to L, then it should remain on L up to the boundary of U, which contradicts that $q_0 \in U - L$. Therefore F_{q_0} takes both positive and negative values, which implies that $\operatorname{rank} F_{q_0} \geq 2$. □

Let $p \in M$ such that $\operatorname{rank} F_p = 1$. Then the subspace

$$W = \{\lambda \in T^*_pM \mid F(\lambda) = 0\}$$

of T^*_pM is an admissible subspace. Let $V_1, \dots, V_n$ be an admissible orthonormal frame on a neighborhood U of p such that $(V_i)_p \in W$, $i \geq 2$, and let N be the maximal integral manifold of the distribution spanned by $V_2, \dots, V_n$ through p. Taking U small we have the following

LEMMA 2.1.6. *$F(\mu) = 0$ and $(X_F)_\mu = 0$ for any $\mu \in T^*_qN$ and $q \in N$. Moreover N is totally geodesic, and F is of rank one on N.*

PROOF. First we consider the case where $p \in M^0$. Around p, F is described as $F = \sum_i f_i V_i^2$. By the definition of V_i we have

$$f_2(p) = \cdots = f_n(p) = 0, \quad f_1(p) \neq 0.$$

Also, in view of the formula $0 = \{E, F\}$, it is easy to see that $\{V_i, f_j\}$ is described as a linear combination of $f_2, \dots, f_n$ with coefficients in $C^\infty(U)$ for any $i, j \geq 2$. Thus we have $f_2 = \cdots = f_n = 0$ and $F = f_1 V_1^2$ on N. This implies that $F(\mu) = 0$ for $\mu \in T^*_qN$, $q \in N$. Taking U small we may assume that $f_1 \neq 0$ on U. Suppose $f_1 > 0$, and take points $q \in N$ and $q_1 \in L_-$. Let

$$\gamma(t) = \pi(\zeta_t\mu), \qquad \mu \in S^*_qM$$

be a geodesic such that $\gamma(t_1) = q_1$. Since $F_q \geq 0$ and $F_{q_1} \leq 0$, it follows that γ is tangent to N and L_- at q and q_1 respectively. Hence $\mu \in S^*_qN$ and $(X_F)_\mu = 0$. Moving q_1 slightly on γ if necessary, we may assume that $q_1 = \gamma(t_1)$ is not conjugate to q along the geodesic γ. Then by the exponential mapping at q, a neighborhood of q_1 in L_- becomes the image of a hypersurface A in T_qM passing through $\sharp(\mu)$. Hence by the semi-definiteness of F on L_- and N we have $A \subset T_qN$, and also $(X_F)_\nu = 0$ for $\nu \in \sharp^{-1}(A)$. Therefore we have

$$(X_F)_\nu = 0 \qquad \text{for all } \nu \in T^*_qN.$$

The fact that N is totally geodesic is easily verified in the same way as the proof of Lemma 2.1.4.

Next we consider the case where $p \in M^s$. Suppose that $\dim \mathcal{F}_p = n - k$. In view of Proposition 1.2.8 there is a basis $F_1, \ldots, F_n$ of $\mathcal{F}$ and a basis $\overline{V}_1, \ldots, \overline{V}_n$ of T_pM such that

$$(F_r)_p = 0, \quad (F_{r+k})_p = \overline{V}^2_{2r+1} + \overline{V}^2_{2r} \quad (1 \leq r \leq k),$$
$$(F_r)_p = \overline{V}^2_r \qquad (2k+1 \leq r \leq n).$$

Since F_p is of rank one, it follows that $2k < n$, and we may assume that $F_p = (F_n)_p$ and $(V_1)_p = \overline{V}_n$. Let I_r $(1 \leq r \leq k)$ be the submanifolds defined before Proposition 1.3.1. We extend $\overline{V}_1, \ldots, \overline{V}_n$ to a semi-admissible frame around p. Then in view of Proposition 1.3.5 $U - \cup_{r=1}^k I_r$ is connected and contained in M^0, and so is $N - \cup_r I_r$. We may also assume that $\operatorname{rank} F_q \geq 1$ and the coefficients of $\overline{V}^2_n$ in F_q is positive for any $q \in U$.

Now, assume that there is a point $q_0 \in N - \cup_r I_r$ such that $\operatorname{rank} F_{q_0} = 1$. Then as shown above, the set of points $q \in N - \cup_r I_r$ such that $\operatorname{rank} F_q = 1$ is open and closed in $N - \cup_r I_r$. Hence it coincides with $N - \cup_r I_r$, and we have

$$F_q = f(q)\overline{V}^2_n, \quad f(q) > 0$$

for any $q \in N$ by continuity.

We shall show the existence of such a point q_0. In the same way as before, we get $(X_F)_\mu = 0$ for any $\mu \in T_p^*N$. Let $p_1 \in I$ be one of the nearest points in I from p. Take points $p_j \in L_-$, $j \geq 2$, so that $p_j \to p_1$ as $j \to \infty$. Then a subsequence of minimal geodesics from p to p_j converges to a minimal geodesic

$$t \mapsto \pi(\zeta_t \lambda), \qquad \lambda \in S_p^*M$$

from p to p_1, and we get $(X_F)_\lambda = 0$. Suppose that $\pi(\zeta_t\lambda) \in U$ for $|t| < \delta$. Then from Lemma 2.1.2,

$$\operatorname{rank} F_{\pi(\zeta_t\lambda)} = 1 \qquad (|t| < \delta).$$

Since $\boldsymbol{R}\overline{V}_n$ is an admissible subspace at each point in U, and since $F_p = f(p)\overline{V}^2_n$, it follows that

$$F_{\pi(\zeta_t\lambda)} = f(\pi(\zeta_t\lambda))\,\overline{V}^2_n \qquad (|t| < \delta).$$

Since $F(\zeta_t\lambda) = 0$, we have $\overline{V}^2_n(\zeta_t\lambda) = 0$, which implies that the geodesic $\pi(\zeta_t\lambda)$ is contained in N. If a point $\notin \cup_r I_r$ is on this geodesic, let it be q_0. If not, move λ slightly so that $\lambda \notin \cup_r T_p^* I_r$. It is possible because, extending or shrinking the geodesic, one can find $t_1 \in \boldsymbol{R}$ such that $\pi(\zeta_{t_1}\lambda) \in L_-$ and $\pi(\zeta_{t_1}\lambda)$ is not conjugate to p. This implies that there is a submanifold A of codimension one in S_p^*M such that the geodesic $\pi(\zeta_t\mu)$ intersects I at some point, and is tangent to L at the point for any $\mu \in A$

In any way we get $\operatorname{rank} F_q = 1$ for any $q \in N$. Then the argument above indicates that $(X_F)_\mu = 0$ and $F(\mu) = 0$ for $\mu \in T^*(N - \cup I_r)$, and that $N - \cup I_r$ is totally geodesic. Hence by continuity the proof is completed. □

LEMMA 2.1.7. *Let p, U, and N be as above. If U is sufficiently small, then*

$$\{q \in U \mid \operatorname{rank} F_q \leq 1\} = N.$$

PROOF. We again assume that $F_p \geq 0$. Let $\pi(\zeta_t\lambda)$, $0 \leq t \leq t_1$, be a geodesic from a point $p_1 \in L_-$ to p, where $\lambda \in S^*_{p_1}L_-$. We may assume that p and p_1 are not mutually conjugate by the reason already stated. Therefore the mapping is a diffeomorphism from a neighborhood U' of $\lambda \in T^*_{p_1}M$ to a neighborhood U_0 of $p \in M$. As we have already seen, this mapping maps $T^*_{p_1}L_- \cap U'$ diffeomorphically into $N \cap U_0$. Hence if U' is sufficiently small, then we have

$$(\pi \circ \zeta_{t_1})^{-1}(U_0 - N) = U' - T^*_{p_1}L_-.$$

This implies that there is $\mu \in T^*_qM$ such that $F_q(\mu) < 0$ for any $q \in U_0 - N$. On the other hand, since $F_p \geq 0$, there is $\nu \in T^*_qM$ such that $F_q(\nu) > 0$ for any $q \in U_0$, provided U_0 small. Hence replacing U with U_0, we have $\operatorname{rank} F_q \geq 2$ for $q \in U - N$. □

LEMMA 2.1.8. *Let $p \notin I$ such that $F_p = 0$. Then the connected component of $\{q \in M \mid F_q = 0\}$ passing through p is also of dimension $n-2$.*

PROOF. Suppose that $\dim \mathcal{F}_p = n - k$, and let $F_1, \dots, F_n$ be as in Proposition 1.2.8. Since $F_p = 0$, there are constants $a_1, \dots, a_k$ such that

$$F = \sum_{i=1}^{k} a_i F_i.$$

In the same way as before, by joining p and points in L_- by geodesics, we have $(X_F)_\mu = 0$ for μ in an $(n-1)$-dimensional subspace of T^*_pM. Hence by Proposition 1.2.8 $a_1, \dots, a_k$ should be zero except one, say a_1. Then $F = a_1F_1$, and the lemma follows from Proposition 1.3.1 and Corollary 1.3.6. □

The lemma above implies that the situation around a point $p \notin I$ such that $F_p = 0$ is completely the same as a point in I. Hence it follows that J is an $(n-1)$-dimensional, closed, totally geodesic submanifold. The connectedness of J is proved as follows: The proof of Lemma 2.1.6 indicates that any point p such that $\operatorname{rank} F_p = 1$ and a point in L are joined by a geodesic in J. Also Lemma 2.1.8 indicates that any point p such that $\operatorname{rank} F_p = 0$ has a connected neighborhood like L in J, which contains a point q such that $\operatorname{rank} F_q = 1$. This completes the proof of Proposition 2.1.1. □

We regard J as a riemannian manifold by means of the induced metric from M. Let $i_1 : T^*J \to T^*M$ be the embedding explained in Preliminary remarks. Put

$$\mathcal{F}' = \{i_1^*F \mid F \in \mathcal{F}\}$$

and

$$\tilde{I} = \{p \in M \mid F_p = 0\}.$$

PROPOSITION 2.1.9. *$(J, \mathcal{F}')$ is a proper Liouville manifold. Moreover, $J^s = J \cap (M^s)'$, where $(M^s)'$ is the union of connected, closed submanifold of codimension 2 that are contained in M^s and not contained in J.*

PROOF. First we shall verify that $\dim \mathcal{F}' = n - 1$ and the conditions (L.1),..., (L.4) are satisfied. We have already seen that $i_1^*F = 0$. Hence $\dim \mathcal{F}' \leq n - 1$. Let $p \in \tilde{I}$ be a point such that $\dim \mathcal{F}_p = n - 1$. Then applying Proposition 1.2.8, we see that there are $F_2, \ldots, F_n \in \mathcal{F}$ such that $i_1^* F_2, \ldots, i_1^* F_n$ are linearly independent at p. Thus

$$\dim \mathcal{F}_p' = \dim \mathcal{F}' = n - 1,$$

and the condition (L.4) is satisfied. Since $E' = i_1^* E$ is the energy function associated with the induced metric, the condition (L.1) is satisfied. Also, (L.3) follows from the fact that $T_q J$ is admissible for any $q \in J - \widetilde{I}$, and that $J - \widetilde{I}$ is dense in J. For the condition (L.2) it is enough to see it on $M^0 \cap J$. Let $V_1, \ldots, V_n$ be an admissible orthonormal frame around $p \in M^0 \cap J$ so that $(F)_p = cV_1^2$. Let $F = G_1, G_2, \ldots, G_n$ be a basis of $\mathcal{F}$. Then G_i are of the form $\sum_j f_{ij} V_j^2$, and

$$i_1^* G_i = \sum_{j \geq 2} f_{ij} V_j^2.$$

Since $\{G_i, G_j\} = 0$ and $\{V_i, V_j\}_p \in \boldsymbol{R} V_i + \boldsymbol{R} V_j$, we get

$$\left\{ \sum_{k \geq 2} f_{ik} V_k^2, \sum_{k \geq 2} f_{jk} V_k^2 \right\} = 0.$$

Hence $(J, \mathcal{F}')$ is a Liouville manifold. Next, we shall show the properness. Let $G \in \mathcal{F}$ such that G and F are mutually linearly independent, and assume that $(i_1^* G)_p = 0$ at some $p \in J$. Suppose first that $\operatorname{rank}(F)_p = 1$. Then there is $c \in \boldsymbol{R}$ such that $(G - cF)_p = 0$. For simplicity we rewrite $G - cF$ as G. Then it follows from Proposition 1.2.8 that $X_G|_{T_p^* M}$ is tangent to $T_p^* J$, and that there is $\mu \in T_p^* J$ such that $(X_G)_\mu \neq 0$. Since the embedding i_1 preserves the symplectic structures, we easily get

$$i_{1*}(X_{i_1^* G})_\mu = (X_G)_\mu \neq 0.$$

Next suppose that $(F)_p = 0$. Then it again follows from Proposition 1.2.8 that $G_p = 0$, and that there is $\mu \in T_p^* J$ such that $(X_G)_\mu$ is non-zero and tangent to $T_p^* J$. Therefore the Liouville manifold $(J, \mathcal{F}')$ is proper. Also, we have seen that $(J)^s = J \cap (M^s)'$. □

2.2. Admissible submanifolds

Let $\mathcal{F}^*$ be the set of all $F \in \mathcal{F} - \{0\}$ such that $\{p \in M \mid F_p = 0\}$ is a closed submanifold of codimension 2. By virtue of Lemma 2.1.8 and Corollary 1.3.6, we have

$$M^s = \cup_{F \in \mathcal{F}^*} \{p \in M \mid F_p = 0\}.$$

We write $F \sim F'$ for $F, F' \in \mathcal{F}^*$ if $F = cF'$ for some $c \in \boldsymbol{R}$. In view of Proposition 1.3.5, it is easily seen that the set $\mathcal{F}^*/\sim$ is at most countable. Let F_m $(m = 1, 2, \ldots)$ be elements of $\mathcal{F}^*$ such that any two of them are mutually linearly independent and $\mathcal{F}^* = \cup_m \boldsymbol{R}^\times F_m$, where $\boldsymbol{R}^\times = \boldsymbol{R} - \{0\}$. Put

$$\begin{aligned} I_m &= \{p \in M \mid (F_m)_p = 0\}, \\ J_m &= \{p \in M \mid \operatorname{rank}(F_m)_p \leq 1\}. \end{aligned}$$

Admissible frames. Fix a point $p_0 \in M^0$ and an admissible orthonormal basis $V_1, \dots, V_n$ of $T_{p_0}M$. Let $c(t)$ $(0 \le t \le 1)$ be a loop in M^0 with the base point p_0. Then along this loop the basis $V_1, \dots, V_n$ is uniquely extended to an admissible orthonormal frame $V_1(c(t)), \dots, V_n(c(t))$. Putting

$$V_i(c(1))^2 = V_{\nu(i)}(c(0))^2 \qquad (1 \le i \le n),$$

we have an element ν of the group $\mathfrak{S}_n$ of permutations of the set $\{1, \dots, n\}$. Clearly this assignment induces the homomorphism

$$\nu : \pi_1(M^0, p_0) \to \mathfrak{S}_n \qquad ([c] \mapsto \nu = \nu([c])),$$

where $\pi_1(M^0, p_0)$ denotes the fundamental group of M^0 with the base point p_0.

PROPOSITION 2.2.1. *Let $\iota : M^0 \to M$ be the inclusion mapping, and let*

$$\iota_* : \pi_1(M^0, p_0) \to \pi_1(M, p_0)$$

be the induced homomorphism between the fundamental groups. Then

$$\mathrm{Ker}\ \iota_* \subset \mathrm{Ker}\ \nu,$$

where Ker *denotes the kernel of the homomorphism.*

Since ι_* is surjective, we have the following

COROLLARY 2.2.2. *There is a finite normal covering $(\widetilde{M}, \widetilde{\mathcal{F}}) \to (M, \mathcal{F})$ with covering transformation group isomorphic to*

$$\pi_1(M^0, p_0)/\mathrm{Ker}\ \nu \quad (< \mathfrak{S}_n)$$

such that the homomorphism

$$\nu : \pi_1(\widetilde{M}^0, \widetilde{p}_0) \to \mathfrak{S}_n$$

is trivial, where $\widetilde{p}_0 \in \widetilde{M}^0$ is a point that covers p_0. Namely, $V_1^2, \dots, V_n^2$ are globally defined on $\widetilde{M}^0$.

PROOF OF PROPOSITION 2.2.1. Let $c_1(t)$ $(0 \le t \le 1)$ be a closed curve in M^0 such that $c_1(0) = c_1(1) = p_0$, and suppose that it represents the trivial element in $\pi_1(M, p_0)$. We shall show that $\nu([c_1]) = 1$. Let $c(s,t)$ $(0 \le s, t \le 1)$ be a mapping into M such that

$$c(1,t) = c_1(t), \quad c(0,t) = c(s,0) = c(s,1) = p_0$$

for any s and t. We may assume that the homotopy mapping c is transversal to every submanifold I_m $(1 \le m \le n-r)$. Hence $c^{-1}(M^s)$ consists of a finite number of interior points, say (s_i, t_i) $1 \le i \le l)$.

Then, by virtue of Proposition 1.3.7 (6), the mappings

$$(s,t) \mapsto (V_j)^2_{c(s,t)} \qquad (1 \le j \le n)$$

are well-defined around each point (s_i, t_i). Thus we have $\nu([c_1]) = 1$. □

Maximal admissible submanifolds. Let $(M, \mathcal{F})$ be a proper Liouville manifold. Fix a point $p_0 \in M^0$, and let W be a k-dimensional admissible subspace of $T_{p_0}M$. We assume that $W \not\subset T_{p_0}J_m$ if $p_0 \in J_m$. Put

$$\mathcal{F}_W = \{F \in \mathcal{F} \mid F^e_{p_0}|_{\sharp^{-1}(W)} = 0\}.$$

Clearly $\dim \mathcal{F}_W = n - k$. Also, put

$$\widehat{W}_p = \{v \in T_pM \mid F^e_p(\sharp^{-1}(v)) = 0 \text{ for any } F \in \mathcal{F}_W\}.$$

It is also clear that $\dim \widehat{W}_p \leq k$ if $p \in M^0$. Now, let N be the set of points $p \in M$ having the following property: There is a C^∞ curve $c(t)$ $(0 \leq t \leq 1)$ such that $c(0) = p_0$, $c(1) = p$, and

$$\dot{c}(t) \in \widehat{W}_{c(t)}, \quad \dim \widehat{W}_{c(t)} = k$$

for any $t \in [0, 1]$. Then we have the following

PROPOSITION 2.2.3. *N is a connected admissible submanifold of dimension k, which is complete with respect to the induced metric from M. Let $\mathcal{F}'$ be the pull-back image cf $\mathcal{F}$ by the embedding $T^*N \to T^*M$. Then $(N, \mathcal{F}')$ is a proper Liouville manifold, and $N^s = N \cap M^s$.*

We shall call N *the maximal admissible submanifold determined by the admissible subspace* W.

PROOF. First we shall consider the case where the homomorphism ν in Proposition 2.2.1 is trivial. In this case there are globally defined sections $V_1^2, \dots, V_n^2$ of S^2TM^0 such that their square roots $V_1, \dots, V_n$ form an admissible basis at every point in M^0. Let W be spanned by $V_1, \dots, V_k$, and let D be the subbundle of TM^0 spanned by $V_1, \dots, V_k$ at each point. We extend D to points on a subset of M^s as follows: Let I be a connected components of some I_m, and let $p \in I$ such that $\dim \mathcal{F}_p = n - 1$. Let $B_m(p)$ be as in Section 1.3. Then clearly $B_m(p) - \{p\} \subset M^0$. Suppose that $T_qB_m(p)$ $(q \neq p)$ is spanned by V_α and V_β, $\alpha < \beta$. Then it is also clear that the numbers α and β do not depend on the points q and p, i.e., they depend only on I. Let I_W be the union of I whose α and β satisfy $\alpha \leq k < \beta$. Then it follows that D can be smoothly extended to $M - I_W$. The extended one is also denote by D. Note that every fibre of D is admissible.

Let N' be the maximal integral manifold of the differential system D on $M - I_W$ through p_0. We shall show that $N' = N$. Let $G_i \in \mathcal{F}_W$ $(k + 1 \leq i \leq n)$ such that $(G_i)_{p_0} = V_i^2$. Then in the same way as the proof of Lemma 2.1.6, we see that

$$(G_i)_p = c_i(p)V_i^2 \qquad (k + 1 \leq i \leq n)$$

for every $p \in N'$. We claim that $c_i(p) \neq 0$ for every $p \in N'$. In fact, assume that $c_i(p) = 0$ at some p. Then $p \in M^s$, and by Proposition 1.2.8 G_i is a linear combination of some $F_m \in \mathcal{F}^*$ with non-zero coefficients. Choose such F_m. Since the fibre D_p at p is admissible, it follows that either $T_p(B_m(p)) \subset D_p$, or $D_p \subset T_pI_m$. If the latter case occurs, then we have $N' \subset I_m$, because I_m is admissible. Since $p_0 \notin I_m$, it is a contradiction. Hence we have $T_p(B_m(p)) \subset D_p$ and thus $B_m(p) \subset N'$. On the other hand, we have $B_m(p) \subset I_{m'}$ for any $m' \neq m$ such that $p \in I_{m'}$, because both submanifolds are admissible. This implies that $(G_i)_q$ is a (non-zero) constant multiple of $(F_m)_q$ for any $q \in B_m(p)$. Since there is $\lambda \in T_q^*B_m(p)$ such that

$F_m(\lambda) \neq 0$ if $q \neq p$ (cf. Lemma 1.3.4 (1)), and since $G_i(\mu) = 0$ for any $\mu \in T^*N'$, we have a contradiction. Hence $c_i(p) \neq 0$ for any $p \in N'$.

From the fact stated above it immediately follows that $T_pN' = \widehat{W}_p$. Hence we have $N' \subset N$. Next, let $p \in N$, and let $c(t)$ $(0 \leq t \leq 1)$ be a curve described in the definition of N. Since $\dim \widehat{W}_{c(t)} = k$, it is easily seen that there is a basis $G'_{n-k}, \dots, G'_n$ of $\mathcal{F}_W$ such that $(G'_i)_{c(t)}$ are of rank one for a fixed t. Let $(G'_i)_{c(t)} \in \boldsymbol{RV}_{m_i}$. Then by continuity we have

$$\{m_{n-k}, \dots, m_n\} = \{n-k, \dots, n\}.$$

This implies that $\dot{c}(t) \in D_{c(t)}$ for every t, and we have $p \in N'$. Therefore we have $N' = N$.

Next we shall show that N is complete with respect to the induced metric. Let d_N and d_M be the distance functions of N and M respectively. Let $\{p_l\} \subset N$ be a Cauchy sequence with respect to d_N. Since $d_M \leq d_N$ generally, $\{p_l\}$ converges to a point $p_\infty \in M$. We shall show that $p_\infty \in N$. If $p_\infty \in M - I_W$, then D is a well-defined differential system on a neighborhood U of p_∞. Since $\{p_l\}$ form a Cauchy sequence of N, they have to lie on a single leaf of the foliation defined by D on U for large l. Therefore we have $p_\infty \in N$ in this case.

Now let us assume that $p_\infty \in I_W$. Then, by the definition of I_W, there is a connected component I of some I_m such that $p_\infty \in I$ and the numbers α and β described above satisfy $\alpha \leq k < \beta$. Since $(G_\beta)_{p_l} = c_\beta(p_l)V_\beta^2$, and since $\boldsymbol{RV}_\beta$ is not admissible at p_∞, we have $(G_\beta)_{p_\infty} = 0$ by continuity. Hence in view of Proposition 1.2.8 we see that G_β is a linear combination of some $F_i \in \mathcal{F}^*$. Let F_i be one of them. Let D_i be the subbundle of TM defined on a neighborhood of p_∞ such that $(D_i)_q = T_q(B_i(p))$ for $q \in B_i(p)$, $p \in I_i$ near p_∞. Then in view of Lemma 1.3.4 (1), (3), we can easily see that $(G^i_\beta)_q$, the $\mathcal{S}^2D_i$-components of G_β, is non-zero for any $q \notin I_i$, and $\operatorname{rank}(G^i_\beta)_q \geq 2$ for $q \notin J_i$ near p_∞.

We claim that $i = m$. In fact, assume that $i \neq m$. Then $V_\beta \notin D_i$. Since $(G_\beta)_q = c_\beta(q)V_\beta^2$ for $q \in N$ near p_∞, it follows that $(G^i_\beta)_q = 0$. This implies that $q \in I_i$ for any $q \in N$ near p_∞. Hence we have $N \subset I_i$, a contradiction. Thus $i = m$.

Therefore we know that G_β is a constant multiple of F_m. Observing G^m_β, we see that $q \in J_m$ for any $q \in N$ near p_∞. Since connected components of $J_m - I_m$ are admissible submanifolds, this indicates that $N \subset J_m$, which contradicts the fact that $W \not\subset T_{p_0}J_m$. Hence $p_\infty \notin I_W$, and the completeness of N follows.

Next, we shall show the properness of $(N, \mathcal{F}')$. Let $\iota_1 : T^*N \to T^*M$ be the inclusion mapping (cf. Preliminary remarks). Let $p \in N^s$ and let $F \in \mathcal{F}$ such that $\iota_1^*F \neq 0$ and $(\iota_1^*F)_p = 0$. Then there are constants a_i $(k+1 \leq i \leq n)$ such that $F - \sum_i a_iG_i$ vanishes at p. In particular, we have $p \in M^s$. For simplicity we rewrite $F - \sum_i a_iG_i$ as F. Then it follows from Proposition 1.2.8 that $X_F|_{T_p^*N}$ is tangent to T_p^*N and there is $\mu \in T_p^*N$ such that $(X_F)_\mu \neq 0$ (note that $(G_i)_p$ are of rank one). Since the embedding ι_1 preserves the symplectic structures, we have

$$\iota_{1*}(X_{\iota_1^*F})_\mu = (X_F)_\mu \neq 0.$$

Thus $(N, \mathcal{F}')$ is proper.

Next, let us show that $N^s = N \cap M^s$. As shown above, $N^s \subset N \cap M^s$. Conversely, let $p \in M^s \cap N$ and suppose that $p \in I_m$. Then we have $B_m(p) \subset N$. Since $F_m|_{T^*(B_m(p))} \neq 0$, we have $F_m|_{T^*N} \neq 0$ in $\mathcal{F}'$. Hence $p \in N^s$.

Finally, let $(M, \mathcal{F})$ be arbitrary proper Liouville manifold. Let

$$\rho : (\widetilde{M}, \widetilde{\mathcal{F}}) \to (M, \mathcal{F})$$

be the finite normal covering described in Corollary 2.2.2. Choose $\widetilde{p}_0 \in \rho^{-1}(p_0)$, and let $\widetilde{N}$ be the maximal admissible submanifold determined by $\widetilde{W} = (\rho_*)^{-1}_{\widetilde{p}_0} W$. Then it is easily seen that $\rho(\widetilde{N}) = N$. Therefore, in order to complete the proposition it suffices to show that $\rho : \widetilde{N} \to N$ is also a covering. Let σ be a covering transformation of $\widetilde{M}$. Since σ preserves each element of $\widetilde{\mathcal{F}}$, it follows that $\sigma(\widetilde{N})$ is the maximal admissible submanifold determined by $\sigma_*(\widetilde{W})$. Now, suppose that $\sigma(\widetilde{p}_0) \in \widetilde{N}$. Since $\mathcal{F}_{\widetilde{W}}$ is preserved by σ, it follows that $\sigma_*(T_{\widetilde{p}_0}\widetilde{N}) = T_{\sigma(\widetilde{p}_0)}\widetilde{N}$. Thus we have $\sigma(\widetilde{N}) = \widetilde{N}$, which implies that $\rho|_{\widetilde{N}}$ is a covering mapping. □

Intersections of admissible submanifolds. Let $(M, \mathcal{F})$ be a proper Liouville manifold such that the homomorphism ν in Proposition 2.2.1 is trivial. Let $V_1^2, \dots, V_n^2$ be sections of $S^2T^*M^0$ so that their square roots $V_1, \dots, V_n$ form an admissible orthonormal frame at each point. Let D (resp. $D^\perp$) be the subbundle of TM^0 spanned by V_1 (resp. $V_2, \dots, V_n$). Let S be a maximal admissible submanifold of $(M, \mathcal{F})$ of codimension 1 such that $TS = D^\perp|_S$. As in the proof of Proposition 2.2.3, we extend D to part of $\cup_m I_m$ where D is well-defined and admissible. Let M' be the set of points where D is defined. Then $M - M'$ is equal to the union of such I that I is a connected component of some I_m and that its normal bundle is continued to an admissible subbundle of TM^0 containing D. Note that $S \subset M'$, because D_p is admissible for every $p \in S$.

For each $p \in M'$ we denote by $C(p)$ the maximal integral curve of D passing through p. Then we have the following

PROPOSITION 2.2.4. *$C(p) \cap S \neq \emptyset$ for any $p \in M'$.*

PROOF. We prove this by induction on $n = \dim M$. The case $n = 2$ is clear by **[Ki1]** Theorem 3.1 and **[IKS]** Theorem 3.1. Now, suppose $n \geq 3$ and assume that the proposition is true for proper Liouville manifolds of dimension less than n.

Let M'' be the set of all $p \in M'$ such that $C(p) \cap S \neq \emptyset$. Since M' is connected, to prove $M'' = M'$ it suffices to show that M'' is open and closed in M'. The openness is clear. To prove the closedness we need a number of lemmas.

Let $G_1, \dots, G_n$ be a basis of $\mathcal{F}$ such that $G_n = 2E$. Describing $\sum_j a_{ij} G_j = V_i^2$ on M^0, we put

$$a_i = \sqrt{\sum_j a_{ij}^2} \qquad (1 \leq i \leq n)$$

and $b_{ij} = a_{ij}/a_i$. Then by Proposition 1.1.3, $V_k b_{ij} = 0$ whenever $k \neq i$, and there is a coordinate system $(x_1, \dots, x_n)$ on a neighborhood of each point of M^0 such that

$$\left(\frac{\partial}{\partial x_i}\right)^2 = \frac{V_i^2}{a_i}.$$

LEMMA 2.2.5. (1) *The function a_1 is smoothly extended to M'.*
(2) *There is a positive C^∞ function A_1 on M' such that $V_1 A_1 = 0$ and $a_1 \geq A_1^2$ everywhere.*

PROOF. We first define A_1 on M^0. Put $B = (b_{ij})$ and let B_{ij} be the matrix obtained by deleting i-th row and j-th column from B. Noticing the formula $\sum_i a_{ij} = \delta_{nj}$, we have

$$a_1 = (-1)^{1+n}\frac{\det B_{1n}}{\det B}.$$

Since $\sum_j b_{ij}^2 = 1$, we also have

$$|\det B| \le \sqrt{\sum_j (\det B_{1j})^2}.$$

Hence, putting

$$A_1 = |\det B_{1n}|^{\frac{1}{2}} \left(\sum_j (\det B_{1j})^2\right)^{-\frac{1}{4}},$$

we have $V_1 A_1 = 0$ and $a_1 \ge A_1^2$.

Next, we show that a_1 and A_1 are smoothly extended to each point $p \in M' - M^0$. Suppose $\dim \mathcal{F}_p = n - k$. By Proposition 1.3.7 we have a coordinate system $(y_1, \dots, y_n)$ on a branched covering space of a neighborhood U of p and functions $\bar{b}_{ij}(y_i)$ such that

$$\sum_j \bar{b}_{ij} G_j = \left(\frac{\partial}{\partial y_i}\right)^2$$

and $(\partial/\partial y_i)^2$ are smooth section of $\mathcal{S}^2TM$ on $U - M^s$. Moreover, the matrix $C = (c_{ij})$ given by

$$c_{ij} = \begin{cases} \bar{b}_{2i-1,j} - \bar{b}_{2i,j} & (1 \le i \le k) \\ (\bar{b}_{2i-1,j} + \bar{b}_{2i,j})/(y_{2i-1}^2 + y_{2i}^2) & (1 \le i \le k) \\ \bar{b}_{ij} & (2k+1 \le i \le n), \end{cases}$$

defines a smooth, non-singular matrix-valued function on U. Also,

$$\left(\frac{\partial}{\partial y_{2i-1}}\right)^2 - \left(\frac{\partial}{\partial y_{2i}}\right)^2, \quad \frac{\left(\frac{\partial}{\partial y_{2i-1}}\right)^2 + \left(\frac{\partial}{\partial y_{2i}}\right)^2}{y_1^2 + y_2^2} \qquad (1 \le i \le k)$$

define smooth sections of $\mathcal{S}^2TU$.

Suppose that $\partial/\partial y_s$ spans D $(s \ge 2k+1)$. Then it is easily seen that

$$a_1 = (-1)^{s+n}\frac{\det C_{sn}}{\det C}, \quad A_1 = |\det C_{sn}|^{\frac{1}{2}} \left(\sum_j (\det C_{sj})^2\right)^{-\frac{1}{4}},$$

and $\det C_{sn} \ne 0$ at p. Hence the lemma follows. □

The lemma above indicates that if an integral curve of D is complete with respect to the length parameter, then it is also complete with respect to the parameter x_1 (it can be defined on each integral curve). Now, we define a subset K of $\cup_m J_m$ as follows: Let $\mathcal{I}$ be the set of subscripts m such that $M - M'$ contains a connected component of I_m. For each $m \in \mathcal{I}$, let K_m be the union of connected components of $J_m - I_m$ such that D is tangent to J_m there. Then, put

$$K = \cup_{m\in\mathcal{I}} K_m \cup (M - M').$$

LEMMA 2.2.6. *Let $p \in M - K(\subset M')$. Then $C(p)$ is complete.*

PROOF. If $p \notin \cup_{m \notin \mathcal{I}} J_m$, then $C(p)$ is a maximal admissible submanifold. Hence it is complete by Proposition 2.2.3. Otherwise, let $m_1, \dots, m_l$ be the set of all $m \notin \mathcal{I}$ such that J_m passes through p. Let L be the connected component of $\cap_{i=1}^{l} J_{m_i}$ containing p. Then one can verify that L is a proper Liouville manifold in the same way as the proof for J_m. Since D is a well-defined subbundle of $T(L \cap M')$, it follows that $C(p)$ becomes a maximal admissible submanifold of L. Hence it is complete. □

LEMMA 2.2.7. *Let $p_0 \in M - K$, and let P be an integral manifold of $D^\perp$ passing through p_0 so that $P \cap K = \emptyset$. Then we have a local diffeomorphism*

$$\Phi : \boldsymbol{R} \times P \to M' \qquad ((x_1, q) \mapsto \Phi(x_1, q))$$

such that $x_1 \mapsto \Phi(x_1, q)$ is an integral curve of D for each $q \in P$, $|\partial\Phi/\partial x_1|^2 = a_1^{-1}$, and the image of $P \ni q \mapsto \Phi(x_1, q)$ is an integral manifold of $D^\perp$ for any $x_1 \in \boldsymbol{R}$.

PROOF. Take a smooth unit normal vector field V_1 along P. Then the mapping Φ is uniquely defined so that $x_1 \mapsto \Phi(x_1, q)$ is an integral curve of D, $|\partial\Phi/\partial x_1|^2 = a_1^{-1}$, and

$$\Phi(0, q) = q, \qquad \frac{\partial \Phi}{\partial x_1}(0, q) = \frac{V_1}{\sqrt{a_1}}.$$

Since $\partial/\partial x_1 = V_1/\sqrt{a_1}$ preserves the distribution $D^\perp$, it follows that the image of $P \ni q \mapsto \Phi(x_1, q)$ is an integral manifold of $D^\perp$. □

We denote by $\#A$ the number of elements of a set A. Using the induction assumption, we have the following

LEMMA 2.2.8. *Let $m \in \mathcal{I}$ and let $p \in K_m \cap M'$. Let $\mathcal{I}'$ be the set of all $m' \in \mathcal{I}$ such that $p \in K_{m'}$. Then:*
1. *$C(p)$ is contained in a connected component of $J_m - I_m$;*
2. *every connected component of I_m is contained in $M - M'$;*
3. *$C(p)$ is not complete;*
4. *there is a boundary point q of $C(p)$ contained in I_m;*
5. *$\#\mathcal{I}' \leq 2$.*

PROOF. Let L be the connected component of $J_m - I_m$ containing p. Assume that $C(p) \not\subset L$. Since D is tangent to L, it follows that $C(p)$ meets a connected component I of the boundary of L in J_m, which is a component of I_m. Since D and I_m are admissible, this implies that $C(p) \subset I_m$, a contradiction. Hence $C(p) \subset L$, and (1) follows.

Now, let us consider the proper Liouville manifold $(J_m, \mathcal{F}')$, where $\mathcal{F}'$ is the pull-back image of $\mathcal{F}$ to $T^* J_m$. Let

$$\rho : (\widetilde{J_m}, \widetilde{\mathcal{F}'}) \to (J_m, \mathcal{F}')$$

be the finite covering described in Corollary 2.2.2. Note that each connected component I of I_m is a maximal admissible submanifold of $(J_m, \mathcal{F}')$. Also, if the normal bundle of I in M is continued to the subbundle of TM^0 spanned by V_i and V_j, then each V_i and V_j are normal to J_m on one of the connected components of $J_m - I_m$ having I as boundary respectively (see Proposition 1.3.7).

Take a component $\widetilde{I}$ of $\rho^{-1}(I_m)$ so that $I = \rho(\widetilde{I}) \subset M - M'$, and let $\widetilde{D}$ be the admissible subbundle of $\widetilde{J}_m^0$ that coincides with the normal bundle of $\widetilde{I}_m = \rho^{-1}(I_m)$ on it. As the case of $D \subset TM$, we extend the defining domain of $\widetilde{D}$ as far as possible. Let us denote it by $\widetilde{J}'_m$. Then by induction assumption any maximal integral curve of $\widetilde{D}$ intersects $\widetilde{I}$. Now, assume that there is a connected component $\widetilde{I}'$ to which $\widetilde{D}$ is tangent. Then any integral curve emanating from a point on $\widetilde{I}'$ is contained in $\widetilde{I}'$, a contradiction. Hence $\widetilde{D}$ is normal to every connected component of $\widetilde{I}_m$, and any maximal integral curve of $\widetilde{D}$ intersects every connected component of $\widetilde{I}_m$.

Let $\widetilde{c}_1(t)$ be a maximal integral curve of $\widetilde{D}$ with the length parameter and put $c_1(t) = \rho(\widetilde{c}_1(t))$. Suppose that $c_1(0) \in K_m$ and $\dot{c}_1(0) \in D$. Suppose also that V_i $(i \neq 1)$ is normal to J_m at $c_1(0)$. Since the curve $c_1(t)$ meets every connected component of I_m, we may assume that there is $t > 0$ such that $c_1(t) \in I_m$. Let $t = t_1 > 0$ be the first positive time such that $c_1(t_1) \in I_m$. Then for a small $\epsilon > 0$, V_i is tangent to the curve and V_1 is normal to J_m at $c_1(t_1 + \epsilon)$. If $t_2 (> t_1)$ is the next time such that $c_1(t_2) \in I_m$, then V_1 is tangent to the curve and V_i is normal to J_m at $c_1(t_2 + \epsilon)$, and so on. This being also true for negative time, we conclude that the normal bundle of every connected component of I_m is continued to the subbundle of TM^0 spanned by V_1 and V_i. Hence (2) follows.

Now, suppose $c_1(0) = p$. Then $c_1(t)$ is an integral curve of D on $[0, t_1)$ and cannot extend further toward the future direction. Thus (3) and (4) follows. Finally, (5) follows from (4) because the number of the end points of a maximal integral curve of D is at most two. $\square$

Let $p_0 \in K_m \cap M'$ $(m \in \mathcal{I})$. Changing the ordering of V_i's if necessary, we may assume that V_2 is normal to J_m at p_0. Let P be a small integral manifold of $D^\perp$ passing through p_0. Since $\boldsymbol{R}V_2$ is admissible on P if P is small enough, we may assume that P is a product manifold $P' \times P''$, where $P'' = P \cap J_m$ and P' is an integral curve of $\boldsymbol{R}V_2$ passing through p_0.

LEMMA 2.2.9. *If P is suitably taken, then there are constants $T > 0$, $\epsilon > 0$, and a mapping*

$$\Psi : [0, 2T] \times (-\epsilon, \epsilon) \times P'' \to M \qquad ((y_1, y_2, q) \mapsto \Psi(y_1, y_2, q))$$

satisfying the following conditions:

1. *$\Psi|_{\{y_1=0\}}$ gives a diffeomorphism from $(-\epsilon, \epsilon) \times P''$ to $P = P' \times P''$ such that it preserves the product structure and $\Psi|_{\{y_1=y_2=0\}}$ is the identity on P'';*
2. *$y_1 \mapsto \Psi(y_1, y_2, q)$ is an integral curve of D on $[0, 2T]$ if $y_2 \neq 0$ and on $[0, T) \cup (T, 2T]$ if $y_2 = 0$;*
3. *$\Psi(y_1, y_2, q) \in I_m$ if and only if $y_1 = T$ and $y_2 = 0$;*
4. *$\Psi(T, y_2, q) \in J_m$ and $\notin K_m$;*
5. *$(y_2, q) \mapsto \Psi(y_1, y_2, q)$ gives a diffeomorphism from P to an integral manifold of $D^\perp$ passing through $\Psi(y_1, 0, p_0)$ if $y_1 \neq T$;*
6. *the norm $|\partial\Psi/\partial y_2|$ is invariant under the transformation $\Psi(y_1, y_2, q) \mapsto \Psi(y_1, -y_2, q)$;*
7. *Ψ is a branched double covering whose covering transformation is given by*

$$(y_1, y_2, q) \mapsto (2T - y_1, -y_2, q).$$

PROOF. By virtue of the previous lemma we know that $C(p_0)$ has an end point $q_0 \in I_m$. Let $p_1 \in C(p_0)$ be a point sufficiently close to q_0. Then, considering the branched covering defined near q_0 described in Proposition 1.3.7, we have constants $T_1, \epsilon_1 > 0$ and a branched double covering

$$\Psi_1 : [0, 2T_1] \times (-\epsilon_1, \epsilon_1) \times P_1'' \to M$$

possessing the similar properties as above. Here, $p_1 = \Psi_1(0, 0, p_1)$ and P is replaced with P_1 = the image of $\Psi|_{\{y_1=0\}}$. Also, I_m is the branched locus on the image of Ψ_1 and represented by $y_1 = T_1$, $y_2 = 0$.

Tracing the proof of Lemma 2.2.7 we also have a diffeomorphism Φ from $[0, T_2] \times P$ $(T > 0)$ into M so that $x_1 \mapsto \Phi(x_1, q)$ is an integral curve of D and $\Phi(0, q) \in P$, $\Phi(T_2, q) \in P_1$ for any $q \in P$. Here, x_1 is the parameter defined before. Then, by changing the parameter x_1 suitably to $y_1 = y_1(x_1)$, the mappings Ψ_1 and Φ are smoothly joined and define a smooth mapping Ψ. The verification of the properties (1), ..., (7) are straightforward except (6), which is shown as follows.

It is enough to consider on the intersection of the image of Ψ and M^0. Let a_{ij} be the functions such that $\sum_j a_{ij} G_j = V_i^2$. Let $\bar{a}_i$ $(i = 1, 2)$ be functions such that $V_i^2 / a_i = (\partial/\partial y_i)^2$. We put $\bar{b}_{ij} = a_{ij}/\bar{a}_i$ $(i = 1, 2)$. Then it is clear that the norm function $\partial/\partial y_i$, which is equal to the coefficient of $(\partial/\partial y_i)^2$ in $2E$, depends only on the functions $\bar{b}_{ij}$ $(i = 1, 2)$ and the ratios

$$a_{i1} : \cdots : a_{in} \qquad (i = 3, \ldots, n).$$

Note that these ratios and the functions $\bar{b}_{2j}$ are invariant on each integral curve of D. Also, the functions $\bar{b}_{1j}$ are invariant under the mapping $y_2 \mapsto -y_2$. By virtue of Proposition 1.3.7 it is verified that the ratios above and the functions $\bar{b}_{ij}$ $(i = 1, 2)$ are invariant under the mapping $y_2 \mapsto -y_2$ on the image of Ψ_1. Therefore (6) follows. □

We now continue the proof of Proposition 2.2.4. Let $p_1, p_2, \ldots$ be a sequence of points in M'' that converges to a point $p_0 \in M'$. Let us show that $p_0 \in M''$. First, suppose that $p_0 \notin K$ and let Φ be the mapping given in Lemma 2.2.7. Since Φ is a local diffeomorphism, it follows that p_k is expressed as $\Phi(s_k, q_k)$ for large k. Then the curve $C(p_k)$ is parametrized as $x_1 \mapsto \Phi(x_1, q_k)$. Let $x_1 = r$ be the point where $C(p_k)$ meets S. Then $\Phi(r, q) \in S$ for any $q \in P$. Since $C(p_0)$ is given by $x_1 \mapsto \Phi(x_1, p_0)$, it therefore follows that $C(p_0) \cap S \neq \emptyset$.

Next, suppose that $p_0 \in K_m$ $(m \in \mathcal{I})$. We first assume that $C(p_0)$ has two ends, say q_0 and q_0'. Then toward these endpoints we have branched coverings Ψ and Ψ' described in Lemma 2.2.9 respectively. By virtue of the property (6) in Lemma 2.2.9 every integral curve of D passing through P forms a closed curve. It intersects P 2 times if q_0' also belongs to I_m, and 4 times if $q_0' \in I_{m'}$, $m \neq m' \in \mathcal{I}$. Since $C(p_k) \cap S \neq \emptyset$, we may assume that $C(p_k)$ meets S on the image of Ψ. Let $\Psi(r, s, q) \in S$. Then $r \neq T$, and S contains the image of $\Psi|_{y_1=r}$. Thus we have $C(p_0) \cap S \neq \emptyset$.

We next assume that $C(p_0)$ has only one end q_0 and S does not meet the image of Ψ. Then toward the converse direction we have a local diffeomorphism

$$\Phi : [0, t_0] \times P \to M \qquad (x_1, q) \mapsto \Psi(x_1, q)$$

such that $x_1 \mapsto \Phi(x_1, q)$ are integral curves of D, where x_1 is the parameter given before. The value t_0 may be chosen arbitrary large, but for doing so it may be necessary that P is replaced with a smaller one. If $p_k \in K$ for every k, then we choose p'_k sufficiently near p_k so that $p'_k \notin K$, $p'_k \in M''$, and p'_k converges to p_0. It is possible because M'' is open. Hence we may assume that all p_k are contained in the same connected component of $M - K$. Then in view of Lemma 2.2.7 there is an integral manifold $\widetilde{P}$ of $D^{\perp}$ contained in a connected component of $P - K$ and the local diffeomorphism

$$\widetilde{\Phi} : \boldsymbol{R} \times \widetilde{P} \to M \qquad (x_1, q) \mapsto \widetilde{\Phi}(x_1, q)$$

so that $\widetilde{\Phi} = \Phi$ on $[0, t_0] \times \widetilde{P}$. Fix k and let $\widetilde{\Phi}(r, q) \in C(p_k) \cap S$. If $r < 0$, then we replace $\{p_k\}$ and $\widetilde{P}$ with their image by the mapping $y_2 \mapsto -y_2$ described in Lemma 2.2.9. So we may assume that $r > 0$. Clearly S contains the image of $\widetilde{\Phi}|_{x_1=r}$. Now we take t_0 so that $t_0 > r$ (and shrink P if necessary). Then, since the image of $\Phi|_{x_1=r}$ is an integral manifold of $D^{\perp}$ and contains points of S, it follows that this image is contained in S. Thus we have $C(p_0) \cap S \neq \emptyset$. This completes the proof of Proposition 2.2.4. □

We also have the following

PROPOSITION 2.2.10. *Let $m \in \mathcal{I}$ and let L be a connected component of $J_m - I_m$ where D is normal to J_m. Then $C(p) \cap L \neq \emptyset$ for any $p \in M' - (K_m \cup I_m)$.*

The proof is similar to that of Proposition 2.2.4, so we omit. By Lemma 2.2.8 (2) and the proof for it we have the following

COROLLARY 2.2.11. *For any $F_m \in \mathcal{F}^*$, the normal bundles of the connected components of I_m are smoothly continued to a single admissible subbundle of TM^0.*

We denote by D_m $(m = 1, 2, \ldots)$ the subbundle of TM^0 of rank 2 that is smoothly continued to the normal bundle of I_m.

PROPOSITION 2.2.12. *$D_m \neq D_l$ if $m \neq l$. In particular, $\#(\mathcal{F}^*/\sim) < \infty$.*

PROOF. Suppose that there are m and l $(m \neq l)$ such that $D_m = D_l$, and suppose that D_m is spanned by V_1 and V_2. Let L_m (resp. L_l) be a connected component of $J_m - I_m$ (resp. $J_l - I_l$) on which V_1 (resp. V_2) is normal. Then by Proposition 2.2.10 we have $L_m \cap L_l \neq \emptyset$. Let $p_0 \in L_m \cap L_l$ such that $p_0 \notin J_k$ for any $k(\neq m, l)$. Let C_i $(i = 1, 2)$ be the maximal integral curve of the subbundle spanned by V_i passing through p_0. Then by Lemma 2.2.8 and the proof for it, the curves C_1 and C_2 have end points $p_1 \in I_l$ and $p_2 \in I_m$ respectively.

Let Q be the maximal admissible submanifold determined by D_m at p_0. Clearly, $C_i \subset Q$, and therefore $p_i \in Q$. Let $\mathcal{F}'$ be the pull-back image of $\mathcal{F}$ by the inclusion $T^*Q \to T^*M$. Since $(Q, \mathcal{F}')$ is a 2-dimensional proper Liouville manifold, we know that $\#(\mathcal{F}'^*/\sim) \leq 1$ by [**Ki1**] and [**IKS**]. Since $F_m|_{T^*Q}$ and $F_l|_{T^*Q}$ vanish at p_2 and p_1 respectively, they are elements of $\mathcal{F}'$. Moreover, they are linearly independent at p_0. Thus we have a contradiction. In particular, the number of elements of $\mathcal{F}^*/\sim$ is bounded by the combination number $n!/2!(n-2)!$. □

2.3. The core of a proper Liouville manifold

Let $(M, \mathcal{F})$ be a proper Liouville manifold. Let $F_m \in \mathcal{F}^*$ $(m = 1, 2, \ldots)$ be as in the beginning of Section 2.2.

LEMMA 2.3.1. *Let $l \neq m$, and let I be a connected component of I_m. Then $J_l \cap I \neq \emptyset$, and at each intersection point their normal spaces are mutually orthogonal. Also, at each intersection point of J_l and J_m their normal spaces are mutually orthogonal.*

PROOF. It suffices to show this proposition under the assumption that the homomorphism ν for $(M, \mathcal{F})$ described in Section 2.2 is trivial. Since $D_l \neq D_m$ by Proposition 2.2.12, there is V_i such that $V_i \in D_l$ and $V_i \notin D_m$. Let $p \in I$ such that $p \notin I_k$ for any $k (\neq m)$, and suppose that $p \notin J_l$. Then by Proposition 2.2.10, the maximal integral curve of the subbundle spanned by V_i passing through p intersects every connected component of $J_l - I_l$ to which V_i is normal. Since this integral curve is contained in I, it follows that $J_l \cap I \neq \emptyset$. Also, Proposition 1.2.8 implies that at each intersection point their normal spaces are mutually orthogonal.

Next, let us show that the normal spaces of J_l and J_m are mutually orthogonal at every $q \in J_l \cap J_m$. If $q \in I_l$, or $\in I_m$, it is clear from Proposition 1.2.8. Suppose that $q \in (J_l - I_l) \cap (J_m - I_m)$. Then $T_q J_l$ and $T_q J_m$ are admissible subspaces. This implies that their normal directions are mutually orthogonal, unless they coincide. If they coincide, then J_l and J_m have to coincide, because these submanifolds are totally geodesic. But this indicates $I_l \subset J_m$, contradicting Proposition 1.2.8. □

We now fix l and consider the proper Liouville manifold $(J_l, \mathcal{F}')$, where $\mathcal{F}'$ is the one induced from $\mathcal{F}$. We put $F'_m = F_m|_{T^* J_l}$.

PROPOSITION 2.3.2. (1) $\mathcal{F}'^* = \cup_{m \neq l} \boldsymbol{R}^\times F'_m$.
(2) *F'_{m_1} and F'_{m_2} are mutually linearly independent if $m_1 \neq m_2$ and $m_1, m_2 \neq l$.*
(3) $\{p \in J_l \mid (F'_m)_p = 0\} = J_l \cap I_m \quad (m \neq l)$.
(4) $\{p \in J_l \mid \mathit{rank}(F'_m)_p \leq 1\} = J_l \cap J_m \quad (m \neq l)$.
(5) $(J_l)^s = \cup_{m \neq l} (I_m \cap J_l)$.
(6) $\#(\mathcal{F}'^*/\sim) = \#(\mathcal{F}^*/\sim) - 1$.

PROOF. First we prove (3). It is clear that $(F'_m)_p = 0$ for any $p \in I_m \cap J_l$. Conversely, assume that $(F'_m)_p = 0$ at $p \in J_l$. Then there is $c \in \boldsymbol{R}$ such that $(F_m - cF_l)_p = 0$. If $(F_m)_p \neq 0$, then we have $p \in J_m$ and $T_p J_m = T_p J_l$. This contradicts Lemma 2.3.1. Hence (3) follows.

Since I_m and J_l intersect transversally, it follows that $I_m \cap J_l$ is a closed submanifold of J_l of codimension 2. Hence $F'_m \in \mathcal{F}'^*$ for $m \neq l$. Similarly, $J_m \cap J_l$ is a submanifold of J_l of codimension 1. Also, it is clear that

$$J_m \cap J_l \subset \{p \in J_l \mid \operatorname{rank}(F'_m)_p \leq 1\}.$$

Since the right-hand side is a connected and closed submanifold of J_l of codimension 1, and since both sides are totally geodesic and complete, it follows that both sides coincide. Thus we have (4). Also, (5) follows from Proposition 2.1.9.

Next we prove (1). We have already seen that $F'_m \in (\mathcal{F}')^*$ for any $m \neq l$. Now let $G \in \mathcal{F}$ such that $G' = G|_{T^* J_l} \in (\mathcal{F}')^*$, and let I' be a connected component of

$$\{p \in J_l \mid G'_p = 0\}.$$

Then in view of (5) we see that I' coincides with a connected component of some $I_m \cap J_l$. This implies that $G' = cF'_m$, $c \in \boldsymbol{R}^\times$. Hence we have (1).

(2) is proved as follows. Assume that $F'_{m_1} = cF'_{m_2}$, $c \neq 0$, $m_1 \neq m_2$. Then we have $J_{m_1} \cap J_l = J_{m_2} \cap J_l$. Let $v \in T_p M$ be a normal vector to J_l at $p \in J_{m_1} \cap J_l$.

Then from Lemma 2.3.1 it follows that $v \in T_pJ_{m_1} \cap T_pJ_{m_2}$. This implies $J_{m_1} = J_{m_2}$, a contradiction. Finally, (6) is an immediate consequence of (1) and (2). □

THEOREM 2.3.3. *$\#(\mathcal{F}^*/\sim)$ is equal to or less than $n-1$. Put $r = n - \#(\mathcal{F}^*/\sim)$ $(1 \le r \le n)$. Then:*

(1) $F_1, \dots, F_{n-r}, E$ are linearly independent;

(2) if $F \in \mathcal{F} - \{0\}$ satisfies $F_p = 0$ at some $p \in M$, then F is of the form

$$F = \sum_{j=1}^{k} a_j F_{m_j}, \qquad a_j \in \boldsymbol{R}^\times, \quad 1 \le m_1 < \cdots < m_k \le n - r,$$

and

$$\{p \in M \mid F_p = 0\} = \cap_{j=1}^{k} I_{m_j}.$$

PROOF. We shall prove the first assertion by induction on $n = \dim M$. If $n = 1$, then $\mathcal{F} = \boldsymbol{R}E$, and $\#(\mathcal{F}^*/\sim) = 0$. Let $n \ge 2$, and assume that the assertion is true for proper Liouville manifolds of dimension less than n. If $\#(\mathcal{F}^*/\sim) = 0$, then there is nothing to prove. If not, by Theorem 2.3.2 (6) we have

$$\#(\mathcal{F}^*/\sim) = \#(\mathcal{F}'^*/\sim) + 1 \le (n-2) + 1 = n - 1.$$

Next we shall show (1). If $r = n$, then there is nothing to prove. Hence we assume $r < n$, and we also assume that $i_1^*F_2, \dots, i_1^*F_{n-r}, i_1^*E$ are linearly independent, where $i_1 : T^*J_1 \to T^*M$ is the inclusion. Suppose that

$$\sum_{m=1}^{n-r} a_m F_m + bE = 0.$$

Then, since $\sum_{m \ge 2} a_m i_1^* F_m + b i_1^* E = 0$, we have $a_2 = \cdots = a_{n-r} = b = 0$. Thus $a_1 F_1 = 0$, and we also have $a_1 = 0$. Therefore the assertion follows by induction on n. (2) is an immediate consequence of Proposition 1.3.5. □

We shall call the the number r in Theorem 2.3.3 the *rank* of the proper Liouville manifold $(M, \mathcal{F})$. The following corollary is clear.

COROLLARY 2.3.4. *The rank of $(J_l, \mathcal{F}')$ is equal to that of $(M, \mathcal{F})$ for any l.*

Applying the procedure of constructing $(J_1, \mathcal{F}')$ out of $(M, \mathcal{F})$ to $(J_1, \mathcal{F}')$, we obtain the proper Liouville manifold $J_1 \cap J_2$ with the induced metric and the induced first integrals. By iterating the procedure successively, we finally get the following

COROLLARY 2.3.5. *$\cap_{m=1}^{n-r} J_m$ is a closed, connected, and totally geodesic submanifold of M, and $I_k \cap \cap_m J_m$ are its closed, admissible submanifolds of codimension one $(1 \le k \le n - r)$. Let $\mathcal{H}$ be the pull-back image of $\mathcal{F}$ by the inclusion mapping $T^*(\cap_m J_m) \to T^*M$. Then $(\cap_m J_m, \mathcal{H})$ is a proper Liouville manifold with the dimension and the rank being r.*

We shall call the Liouville manifold $(\cap_m J_m, \mathcal{H})$ the core submanifold of $(M, \mathcal{F})$. Let W_l be a unit normal vector field to J_l. Though it is not necessarily globally defined, its square W_l^2 is well-defined. Then each F_l $(1 \le l \le n - r)$ is written as

$$(F_l)_p = f_l(p) W_l^2$$

for $p \in \bigcap_m J_m$, and $W_1, \ldots, W_{n-r}$ form a basis of $N_p(\bigcap_m J_m)$. We say that two (not identically zero) functions on a manifold are equivalent if the ratio is constant. Let $[f]$ denote the equivalence class containing the function f. Then we shall call the pair

$$((\bigcap_m J_m, \mathcal{H}), \{[f_1], \ldots, [f_{n-r}]\})$$

of the Liouville manifold of the maximal rank and the set of the $n-r$ equivalence classes of functions on it the *core* of the Liouville manifold $(M, \mathcal{F})$.

Let P^1 and P^2 be two r-dimensional proper Liouville manifolds of rank r, and let $f_1^i, \ldots, f_{n-r}^i$ be functions on P^i $(i = 1, 2;\ n \geq r)$. Then we shall say that the two pairs

$$(P^i, \{[f_1^i], \ldots, [f_{n-r}^i]\}) \qquad (i = 1, 2)$$

are mutually *isomorphic* if there is an isomorphism $\phi : P^1 \to P^2$ of Liouville manifolds such that

$$\{\, [\phi^* f_j^2] \mid j = 1, \ldots, n-r \} = \{\, [f_j^1] \mid j = 1, \ldots, n-r \}.$$

Clearly the cores of mutually isomorphic Liouville manifolds are mutually isomorphic. In later sections we shall observe that the converse is also true for Liouville manifolds of rank one.

3. Proper Liouville Manifolds of Rank One

3.1. Configuration of zeros and type of cores

In Sections 3.1 and 3.2 we shall investigate the conditions that will characterize the cores of Liouville manifolds of rank one. Let $(M, \mathcal{F})$ be an n-dimensional proper Liouville manifold of rank one $(n \geq 2)$. Let F_k, I_k, J_k, and $f_k \in C^\infty(\bigcap_m J_m)$ $(1 \leq m \leq n-1)$ be as in Section 2.3. By Corollary 2.3.5, the core submanifold $\bigcap_{m=1}^{n-1} J_m$ in this case is nothing but a connected one-dimensional riemannian manifold, and $I_k \cap \bigcap_m J_m = f_k^{-1}(0)$ are discrete sets of points on it. In this section we shall observe the configuration of these points. For this purpose we first review the results for 2-dimensional case obtained in **[Kol1]**, **[Ki1]**, and **[IKS]**. Note that 2-dimensional proper Liouville manifold of rank one corresponds to complete Liouville surface with $\mathcal{N} (= M^s) \neq \emptyset$ in **[Ki1]** and **[IKS]**. In this case we write I and J instead of I_1 and J_1 respectively.

Proposition 3.1.1. *When $n = 2$, $M^s = I$ consists of finitely many points. There are four cases:*

(A) *$M \simeq S^2$ (diffeomorphic), $\#I = 4$, $J \simeq S^1$, I is invariant under the antipodal mapping of J;*
(B) *$M \simeq \boldsymbol{R}P^2$, $\#I = 2$, $J \simeq S^1$;*
(C) *$M \simeq \boldsymbol{R}^2$, $\#I = 2$, $J \simeq \boldsymbol{R}$;*
(D) *$M \simeq \boldsymbol{R}^2$, $\#I = 1$, $J \simeq \boldsymbol{R}$.*

In every case except (B) *there is a branched double covering $\Phi : R \to M$ possessing the following properties:*

(1) *$R = (\boldsymbol{R}/\alpha_1\boldsymbol{Z}) \times (\boldsymbol{R}/\alpha_2\boldsymbol{Z})$ for the case* (A), *$R = (\boldsymbol{R}/\alpha_1\boldsymbol{Z}) \times (-\beta_2, \beta_2)$ for the case* (C), *and $R = (-\beta_1, \beta_1) \times (-\beta_2, \beta_2)$ for the case* (D), *where α_1, α_2 are positive real numbers and β_1, β_2 are either positive real numbers or ∞;*
(2) *the inverse image of any point in M is of the form $\{x, -x\}$, where $x = (x_1, x_2)$ is the natural coordinates on R;*

(3) *the branch locus* $\Phi(\{x \in R \mid x = -x\})$ *is equal to* I;
(4) $\Phi^{-1}(J)$ *is equal to* $\{x_1 = 0, \alpha_1/2 \text{ or } x_2 = 0, \alpha_2/2\}$ *for the case* (A), *to* $\{x_1 = 0, \alpha_1/2 \text{ or } x_2 = 0\}$ *for the case* (C), *and to* $\{x_1 = 0 \text{ or } x_2 = 0\}$ *for the case* (D);
(5) *the* Φ*-image of* $\{x_i = \text{constant}\}$ $(i = 1, 2)$ *are maximal admissible submanifolds of* M *unless they are part of* J;
(6) *there are* C^∞ *functions* $f_1(x_1)$ *and* $f_2(x_2)$ *such that*

$$F = \Phi_* \left((f_1 - f_2)^{-1} \left(f_2 \left(\frac{\partial}{\partial x_1} \right)^2 + f_1 \left(\frac{\partial}{\partial x_2} \right)^2 \right) \right),$$

$$2E = \Phi_* \left((f_1 - f_2)^{-1} \left(\left(\frac{\partial}{\partial x_1} \right)^2 + \left(\frac{\partial}{\partial x_2} \right)^2 \right) \right),$$

where F *is a suitable element of* $\mathcal{F}^*$.

For simplicity we shall call 1-dimensional maximal admissible submanifolds *coordinate lines*. We need to know how these submanifolds and J intersect to one another. By the proposition above one can easily see the following facts. In every case described in the proposition above, coordinate lines are closed submanifolds diffeomorphic to the circle S^1 or the real line $\boldsymbol{R}$. In the case (A) every coordinate line is diffeomorphic to S^1, and intersects J at two points. Those two points belong to mutually opposite connected components of $J - I$ respectively, and have the same distance from each of the remaining two components. Two coordinate lines intersect orthogonally at two points if they pass distinct pairs of components of $J - I$, and they do not intersect if they pass the same pair of components. In the case (C) both kinds of coordinate lines exist. Any coordinate line diffeomorphic to S^1 intersects J at two points. Those two points belong to mutually different unbounded components of $J - I$ respectively, and have the same distance from the unique bounded component. Any coordinate line diffeomorphic to $\boldsymbol{R}$ intersects J at one point that belongs to the bounded component of $J - I$. Two coordinate lines of mutually different kinds intersect orthogonally at two points, and those of the same kind do not intersect. In the case (D) every coordinate line is diffeomorphic to $\boldsymbol{R}$ and intersects J at one point. Two coordinate lines passing through mutually different connected components of $J - I$ intersect orthogonally at two points, and those passing through the same component do not intersect.

We shall say that a 2-dimensional proper Liouville manifold of rank one is of type (A), etc. if the corresponding case occurs.

LEMMA 3.1.2. *When* $n = 3$, *the 2-dimensional Liouville manifolds* J_1 *and* J_2 *are of the same type. Identifying* $J_1 \cap J_2$ *with* $\boldsymbol{R}/l\boldsymbol{Z}$ $(l > 0)$ *or* $\boldsymbol{R}$ *isometrically in a suitable way, we have constants* β_1 *and* β_2 *so that* $I_m \cap J_1 \cap J_2$ *are described as follows.*

(1) $J_1 \simeq J_2 \simeq S^2$, $J_1 \cap J_2 = \boldsymbol{R}/l\boldsymbol{Z}$. $0 < \beta_1 < \beta_2 < l/4$.

$$I_m \cap J_1 \cap J_2 = \{\pm\beta_m, \ \frac{l}{2} \pm \beta_m\} \qquad (m = 1, 2).$$

The numbers of the connected components of I_1 *and* I_2 *are both two.*

(2) $J_1 \simeq J_2 \simeq \boldsymbol{R}P^2$, $J_1 \cap J_2 = \boldsymbol{R}/l\boldsymbol{Z}$. $0 < \beta_1 < \beta_2 < l/2$.

$$I_m \cap J_1 \cap J_2 = \{\pm\beta_m\} \qquad (m = 1, 2).$$

I_1 and I_2 are connected.

(3) $J_1 \simeq J_2 \simeq \boldsymbol{R}^2$, $J_1 \cap J_2 = \boldsymbol{R}$. $0 < \beta_1 < \beta_2$. *By exchanging I_1 and I_2 (and also J_1 and J_2) if necessary,*

$$I_m \cap J_1 \cap J_2 = \{\pm\beta_m\} \qquad (m = 1, 2).$$

I_1 has two connected components, and I_2 is connected.

(4) $J_1 \simeq J_2 \simeq \boldsymbol{R}^2$, $J_1 \cap J_2 = \boldsymbol{R}$. $\beta_1 < \beta_2$.

$$I_m \cap J_1 \cap J_2 = \{\beta_m\} \qquad (m = 1, 2).$$

I_1 and I_2 are connected.

REMARK. In (1) and (2) the numbers in the braces should be considered modulo $l\boldsymbol{Z}$.

PROOF. First of all, note that $I_1 \cap I_2 = \emptyset$. (If I_1 and I_2 intersect, then they intersect transversally. This contradicts that $n = 3$.) We first consider the case where $J_1 \cap J_2$ is diffeomorphic to S^1. Since $J_1 \cap J_2$ is also the core submanifold of the Liouville surfaces J_1 and J_2, it follows that J_1 and J_2 are diffeomorphic to S^2 or $\boldsymbol{RP}^2$. Assume that both J_1 and J_2 are diffeomorphic to S^2. Applying the results above to Liouville surfaces J_m, we have

$$\#(J_1 \cap I_2) = \#(J_2 \cap I_1) = 4.$$

Clearly both sets are invariant under the antipodal mapping on $J_1 \cap J_2$. Since each connected component of I_2 is a coordinate line of the Liouville surface J_2, it intersects $J_1 \cap J_2$ at two points. Therefore I_2 has two connected components, and so does I_1. Also, these two points have the same distance from a connected component of $J_1 \cap J_2 - I_1$ to which these points do not belong. All of these facts indicate that the case (1) occurs here.

If both J_1 and J_2 are diffeomorphic to $\boldsymbol{RP}^2$, then a similar argument as above shows that the case (2) occurs. Next, assume that J_1 is diffeomorphic to S^2 and J_2 is diffeomorphic to $\boldsymbol{RP}^2$. In this case we see that

$$\#(J_1 \cap I_2) = 4, \quad \#(J_2 \cap I_1) = 2,$$

and $J_1 \cap I_2$ is invariant under the antipodal mapping on $J_1 \cap J_2$. Since I_1 is a disjoint union of coordinate lines on the Liouville surface J_1, it thus follows that I_1 is connected. Also, the two points in $J_2 \cap I_1$ should belong to mutually opposite connected components of $J_1 \cap J_2 - I_2$. Let $\tilde{J}_2$ be the double covering of the Liouville surface J_2, and $\tilde{I}_2$ the inverse image of I_2 by the covering mapping. Then it is clear that $\tilde{I}_2$ intersects every connected component of the inverse image of $J_1 \cap J_2 - I_1$. This implies that $\tilde{I}_2$ has self-intersection, and so does I_2. Since I_2 is a submanifold, it is a contradiction. Therefore this case does not occur.

Next we shall consider the case where $J_1 \cap J_2$ is diffeomorphic to $\boldsymbol{R}$. In this case we see that both J_1 and J_2 are diffeomorphic to $\boldsymbol{R}^2$ by the results for 2-dimensional case. Assume that $J_1 \cap I_2$ and $J_2 \cap I_1$ consist of one point and two points respectively. Then the Liouville surface J_1 is of type (D). Also, since I_1 has no self-intersection, the two points in $J_2 \cap I_1$ lie on the same connected component of $J_1 \cap J_2 - I_2$. Thus, the Liouville surface J_2 being of type (C), the coordinate line

I_2 becomes a circle. This implies that $\#(J_1 \cap I_2) = 2$, a contradiction. Therefore we have

$$\#(J_1 \cap I_2) = \#(J_2 \cap I_1) \quad (= 1 \text{ or } 2).$$

If this number is 1, then the case (4) occurs. If it is 2, then observing the configuration of coordinate lines, we see that the case (3) occurs. □

PROPOSITION 3.1.3. *Assume $n \geq 3$. Then the 2-dimensional proper Liouville manifolds $\cap_{k\neq m} J_k$ $(1 \leq m \leq n-1)$ are of the same type. Changing the order of J_m (and accordingly that of I_m) if necessary, and identifying $\cap_k J_k$ with $\boldsymbol{R}/l\boldsymbol{Z}$ $(l > 0)$ or $\boldsymbol{R}$ isometrically in a suitable way, we have constants $\beta_1, \dots, \beta_{n-1}$ such that $I_m \cap \cap_k J_k$ are described as in the following list.*

(1) *$\cap_{k\neq m} J_k$ are of type* (A): $0 < \beta_1 < \cdots < \beta_{n-1} < l/4$,

$$I_m \cap \cap_k J_k = \{\pm\beta_m, \frac{l}{2} \pm \beta_m\} \qquad (1 \leq m \leq n-1).$$

(2) *$\cap_{k\neq m} J_k$ are of type* (B): $0 < \beta_1 < \cdots < \beta_{n-1} < l/2$,

$$I_m \cap \cap_k J_k = \{\pm\beta_m\} \qquad (1 \leq m \leq n-1).$$

(3) *$\cap_{k\neq m} J_k$ are of type* (C): $0 < \beta_1 < \cdots < \beta_{n-1}$,

$$I_m \cap \cap_k J_k = \{\pm\beta_m\} \qquad (1 \leq m \leq n-1).$$

(4) *$\cap_{k\neq m} J_k$ are of type* (D): $\beta_1 < \cdots < \beta_{n-1}$,

$$I_m \cap \cap_k J_k = \{\beta_m\} \qquad (1 \leq m \leq n-1).$$

PROOF. We may assume that $n \geq 4$. Since $\cap_{k\neq 1,m} J_k$ is a 3-dimensional proper Liouville manifold of rank one for any $m \geq 2$, it follows from the previous lemma that the 2-dimensional Liouville manifolds $\cap_{k\neq 1} J_k$ and $\cap_{k\neq m} J_k$ are of the same type. Hence the first assertion has been proved.

We now assume that $\cap_{k\neq m} J_k$ are of type (A). Applying the previous lemma to the 3-dimensional Liouville manifold $\cap_{k\neq j,m} J_k$ $(j \neq m)$, we see that the sets $I_j \cap \cap_k J_k$ and $I_m \cap_k J_k$ have no common point. Let l be the length of the circle $\cap_k J_k$. Then we can identify $\cap_k J_k$ isometrically with $\boldsymbol{R}/l\boldsymbol{Z}$ by choosing the midpoint of two points in $I_1 \cap \cap_k J_k$ that are not mutually antipodal, and by choosing an orientation arbitrary. Applying the previous lemma to $\cap_{k\neq 1,m} J_k$, we get $n-1$ constants $\beta_1, \dots, \beta_{n-1}$ such that

$$I_m \cap \cap_k J_k = \{\pm\beta_m, \frac{l}{2} \pm \beta_m\} \qquad (1 \leq m \leq n-1).$$

Hence changing the order so that the ordering

$$0 < \beta_1 < \cdots < \beta_{n-1} < \frac{l}{4}$$

holds, we have the case (1). The other cases are also proved in the same way. □

We shall define the type of a rank one, proper Liouville manifold $(M, \mathcal{F})$ as the type of 2-dimensional Liouville manifolds $\cap_{k\neq m} J_k$.

PROPOSITION 3.1.4. *If* $(M, \mathcal{F})$ *is of type* (A) *or* (C) *or* (D), *then* M *is simply connected. If it is of type* (B), *then* $\#\pi_1(M) = 2$, *and the universal covering of* $(M, \mathcal{F})$ *is of type* (A).

PROOF. Let $(\widetilde{M}, \widetilde{\mathcal{F}})$ be the universal covering of $(M, \mathcal{F})$, and let $\rho : \widetilde{M} \to M$ be the covering mapping. Since

$$\widetilde{J}_k = \rho^{-1}(J_k)$$

are just the corresponding objects of $(\widetilde{M}, \widetilde{\mathcal{F}})$, it follows that $\cap_{k \neq m} \widetilde{J}_k$ are connected, 2-dimensional proper Liouville manifolds of rank one and are preserved by the covering transformation group G. Therefore we have the normal covering

$$\rho : \cap_{k \neq m} \widetilde{J}_k \to \cap_{k \neq m} J_k$$

whose covering transformation group is also G. If $(M, \mathcal{F})$ is of type (A) or (C) or (D), then $\cap_{k \neq m} J_k$ is simply connected. Hence $G = \{1\}$ in these cases. If it is of type (B), then $\cap_{k \neq m} J_k$ is diffeomorphic to $\boldsymbol{R}P^2$. On the other hand, since $\widetilde{M}$ is simply connected, it follows that the normal bundles of $\widetilde{J}_k$ are trivial. This implies that $\cap_{k \neq m} \widetilde{J}_k$ is orientable, and we have $\#G = 2$ in this case. □

3.2. Possible cores

Let $(M, \mathcal{F})$ be a proper Liouville manifold of rank one. We shall identify $\cap_k J_k$ with $\boldsymbol{R}/l\boldsymbol{Z}$ or $\boldsymbol{R}$ and adopt the ordering of the representatives F_m $(1 \leq m \leq n-1)$ of elements of $\mathcal{F}^*/\sim$ so that each $I_m \cap \cap_k J_k$ becomes as described in Proposition 3.1.3. Let $f_1, \ldots, f_{n-1}$ be functions on $\cap_k J_k$ given in Section 2.3. Under the identification f_m are regarded as C^∞ functions on $\boldsymbol{R}/l\boldsymbol{Z}$ or $\boldsymbol{R}$.

We shall first consider the case where $(M, \mathcal{F})$ is of type (A). Let t be the canonical coordinate function on $\boldsymbol{R}$. For simplicity we shall express points and subsets of $\boldsymbol{R}/l\boldsymbol{Z}$ by using the coordinate on $\boldsymbol{R}$. For instance,

$$\{t \in \boldsymbol{R}/l\boldsymbol{Z} \mid a \leq t \leq b\}$$

expresses the subset of $\boldsymbol{R}/l\boldsymbol{Z}$ that is the image of the interval $[a, b]$ in $\boldsymbol{R}$ by the covering mapping $\boldsymbol{R} \to \boldsymbol{R}/l\boldsymbol{Z}$. Let $\beta_1, \ldots, \beta_{n-1}$ be as in Proposition 3.1.3.

PROPOSITION 3.2.1. *Replacing* f_m *with their non-zero constant multiples if necessary, we get:*

(A.0)
$$0 < \beta_1 < \cdots < \beta_{n-1} < l/4$$

(A.1)
$$f_m(t) \begin{cases} = 0 & \text{if} \quad t = \pm\beta_m, \ \frac{l}{2} \pm \beta_m, \\ > 0 & \text{if} \quad -\beta_m < t < \beta_m, \\ < 0 & \text{if} \quad \beta_m < t < \frac{l}{2} - \beta_m; \end{cases}$$

(A.2)
$$f_m{}'(\beta_m) < 0;$$

(A.3)
$$\begin{aligned} f_m(t) &= f_m(-t) && \text{if} \quad \beta_1 \leq t \leq \frac{l}{2} - \beta_1, \\ f_m(t) &= f_m(\frac{l}{2} - t) && \text{if} \quad -\beta_{n-1} \leq t \leq \beta_{n-1}; \end{aligned}$$

(A.4) $$f_1(t) < \cdots < f_{n-1}(t) \quad \textit{for any} \quad t \in \boldsymbol{R}/l\boldsymbol{Z}.$$

PROOF. (A.0), (A.1) are restatements of Proposition 3.1.3 (1). Also, (A.2) easily follows from the proof of Lemma 2.1.3. Now we shall prove (A.3). Let $t_0 \in \boldsymbol{R}/l\boldsymbol{Z}$ such that $\beta_1 < t_0 < l/2 - \beta_1$ and $t_0 \neq \beta_2, \ldots, \beta_{n-1}$, and let W be the $(n-1)$-dimensional subspace of $T_{t_0}M$ that is orthogonal to $T_{t_0}(\boldsymbol{R}/l\boldsymbol{Z})$. Let N be the maximal admissible submanifold determined by W. Since $N \cap \bigcap_{m\neq 1} J_m$ is a union of coordinate lines on the 2-dimensional Liouville manifold $\bigcap_{m\neq 1} J_m$, it passes through $-t_0 \in \boldsymbol{R}/l\boldsymbol{Z}$. From the definition of N, we get

$$-\sum_{m=1}^{n-1} \frac{1}{f_m(t_0)} F_m + 2E \in \mathcal{F}_W.$$

This implies that

$$-\sum_{m=1}^{n-1} \frac{1}{f_m(t_0)} (F_m)_{-t_0} + 2E_{-t_0} = \left(\frac{\partial}{\partial t}\right)^2.$$

Since $(F_m)_{-t_0}$ is equal to $f_m(-t_0)$ times the square of a unit vector, it therefore follows that $f_m(t_0) = f_m(-t_0)$. The other one is also proved in a similar way.

Next we shall prove (A.4). We assume that (A.1) is already satisfied. Fix k $(1 \leq k \leq n-2)$, and let t_1 and t_2 be two points on $\boldsymbol{R}/l\boldsymbol{Z}$ such that $-\beta_k < t_1 < \beta_k$ and $\beta_{k+1} < t_2 < l/2 - \beta_{k+1}$. Let L_1 and L_2 be coordinate lines on the 2-dimensional Liouville manifold $\bigcap_{m\neq k} J_m$ passing through t_1 and t_2 respectively. First, assume that $t_1, t_2 \neq \pm\beta_m, l/2 \pm \beta_m$ for any m. Then on each L_i we have

$$-\sum_{m=1}^{n-1} \frac{1}{f_m(t_i)} F_m + 2E = W_i^2,$$

where W_i is a non-zero vector field along L_i which is tangent to $\bigcap_{m\neq k} J_m$ and normal to L_i. Since $t_1 < \beta_k < t_2$, L_1 and L_2 intersects at two points. Let q be one of these points. Considering the formula above at q, we have

$$-\sum_{\substack{1\leq m\leq n-1 \\ m\neq k}} \left(\frac{f_k(t_1)}{f_m(t_1)} - \frac{f_k(t_2)}{f_m(t_2)}\right) F_m + (f_k(t_1) - f_k(t_2))2E = f_k(t_1)W_1^2 - f_k(t_2)W_2^2.$$

Note that F_m $(m \neq k)$ are described in the form $h_m U_m^2$ on $\bigcap_{l\neq k} J_l$, where U_m $(m \neq k)$ is an orthonormal frame of the normal bundle of $\bigcap_{l\neq k} J_l$. Hence we have

$$\left(\frac{f_k(t_1)}{f_{k+1}(t_1)} - \frac{f_k(t_2)}{f_{k+1}(t_2)}\right) h_{k+1}(q) = f_k(t_1) - f_k(t_2).$$

Since the function h_{k+1} is continuous, it clearly follows that the formula above holds for any t_1 $(-\beta_k < t_1 < \beta_k)$ and t_2 $(\beta_{k+1} < t_2 < l/2 - \beta_{k+1})$. In this case one can easily see that $h_{k+1}(q) < 0$. Since $f_k(t_1) > 0 > f_k(t_2)$, we therefore get

$$\frac{f_k(t_1)}{f_{k+1}(t_1)} < \frac{f_k(t_2)}{f_{k+1}(t_2)}.$$

Clearly the left-hand side of the inequality above takes its maximum in the interval $-\beta_k < t_1 < \beta_k$, and also the right-hand side takes its minimum in $\beta_{k+1} < t_2 < l/2 - \beta_{k+1}$. Hence there is a constant $c_k > 0$ such that

$$\max_{-\beta_k < t_1 < \beta_k} \frac{f_k(t_1)}{c_k f_{k+1}(t_1)} < 1 < \min_{\beta_{k+1} < t_2 < l/2 - \beta_{k+1}} \frac{f_k(t_2)}{c_k f_{k+1}(t_2)}.$$

Therefore, replacing F_k with $c_1 \ldots c_{k-1} F_k$, we have (A.4). □

Similarly, we have the following proposition for other types.

PROPOSITION 3.2.2. *Let* $(M, \mathcal{F})$ *be of type* (B) *or* (C) *or* (D). *Replacing* F_m *with their non-zero constant multiples if necessary, we get the following properties.*
(1) *Case of type* (B):

(B.0)
$$0 < \beta_1 < \cdots < \beta_{n-1} < l/2$$

(B.1)
$$f_m(t) \begin{cases} = 0 & \text{if} \quad t = \pm\beta_m, \\ > 0 & \text{if} \quad -\beta_m < t < \beta_m, \\ < 0 & \text{if} \quad \beta_m < t < l - \beta_m; \end{cases}$$

(B.2)
$$f_m{}'(\beta_m) < 0;$$

(B.3)
$$f_m(t) = f_m(-t) \quad \text{for any} \quad t \in \boldsymbol{R}/l\boldsymbol{Z};$$

(B.4)
$$f_1(t) < \cdots < f_{n-1}(t) \quad \text{for any} \quad t \in \boldsymbol{R}/l\boldsymbol{Z}.$$

(2) *Case of type* (C):

(C.0)
$$0 < \beta_1 < \cdots < \beta_{n-1}$$

(C.1)
$$f_m(t) \begin{cases} = 0 & \text{if} \quad t = \pm\beta_m, \\ > 0 & \text{if} \quad |t| < \beta_m, \\ < 0 & \text{if} \quad |t| > \beta_m; \end{cases}$$

(C.2)
$$f_m{}'(\beta_m) < 0;$$

(C.3)
$$f_m(t) = f_m(-t) \quad \text{if} \quad |t| \geq \beta_1;$$

(C.4)
$$f_1(t) < \cdots < f_{n-1}(t) \quad \text{for any} \quad t \in \boldsymbol{R}.$$

(3) *Case of type* (D):

(D.0)
$$\beta_1 < \cdots < \beta_{n-1}$$

(D.1)
$$f_m(t) \begin{cases} = 0 & \text{if} \quad t = \beta_m, \\ > 0 & \text{if} \quad t < \beta_m, \\ < 0 & \text{if} \quad t > \beta_m; \end{cases}$$

(D.2)
$$f_m{}'(\beta_m) < 0;$$

(D.3) $$f_1(t) < \cdots < f_{n-1}(t) \quad \textit{for any} \quad t \in \boldsymbol{R}.$$

PROOF. Everything is proved in the same way as the proof of the previous proposition except the property (B.3). So we shall give a proof only for it. In view of Proposition 3.1.4, the universal (double) covering of $(M, \mathcal{F})$ is of type (A), and the core is also covered by the core of the covering manifold, which is identified with $\boldsymbol{R}/2l\boldsymbol{Z}$. Let ρ denote the covering mapping. Since the functions $\tilde{f}_m(t) = f_m(\rho(t))$ are invariant under the covering transformation $t \mapsto l + t$, we have

$$\tilde{f}_m(t) = \tilde{f}_m(l-t) = \tilde{f}_m(-t) \qquad \text{for any } t \in \boldsymbol{R}/2l\boldsymbol{Z}$$

instead of (A.3). Therefore (B.3) follows. □

3.3. Constructing a Liouville manifold from a possible core

Let X stand for one of A, B, C, and D, and correspondingly, let P be $\boldsymbol{R}/l\boldsymbol{Z}$ or $\boldsymbol{R}$. Here l is an arbitrary positive real number. Let $f_1, \ldots, f_{n-1}$ be functions on P satisfying the conditions (X.1),..., (X.4) (if X = D, then (D.1), (D.2), and (D.3)) for some constants $\beta_1, \ldots, \beta_{n-1}$ satisfying the condition (X.0) The pair

$$(P, \{[f_1], \ldots, [f_{n-1}]\})$$

of P and the set of equivalence classes of functions so obtained will be called a *possible core* (of type (X)). In this section we shall prove the following

THEOREM 3.3.1. *For each possible core of type* (X), *a proper Liouville manifold of rank one and type* (X) *is naturally constructed so that its core is isomorphic to the given one. Furthermore, the constructed manifold is diffeomorphic to* S^n *if* X = A, $\boldsymbol{R}P^n$ *if* X = B, *and* $\boldsymbol{R}^n$ *if* X = C *or* D.

Let $(P, \{[f_1], \ldots, [f_{n-1}]\})$ be a possible core of type (X), and let $\beta_1, \ldots, \beta_{n-1}$ be the corresponding constants. We first consider the case where X = A.

As the first step, we shall construct a torus $R = \prod_{i=1}^n (\boldsymbol{R}/\alpha_i \boldsymbol{Z})$. For simplicity we put $\beta_0 = -\beta_1$, $\beta_n = l/2 - \beta_{n-1}$ $(\beta_0 < \beta_1 < \cdots < \beta_{n-1} < \beta_n)$. Define positive numbers $\alpha_1, \ldots, \alpha_n$ by

$$\int_{\beta_{i-1}}^{\beta_i} \frac{dt}{\sqrt{(-1)^{i-1} f_1(t) \ldots f_{n-1}(t)}} = \begin{cases} \frac{\alpha_i}{4} & (2 \le i \le n-1) \\ \frac{\alpha_i}{2} & (i = 1, n). \end{cases}$$

Putting $t = t_i$ on the interval $[\beta_{i-1}, \beta_i]$, we define the mapping $t_i \mapsto x_i$,

$$[\beta_{i-1}, \beta_i] \to \begin{cases} [0, \frac{\alpha_1}{2}] & (i = 1) \\ [0, \frac{\alpha_i}{4}] & (2 \le i \le n-1) \\ [-\frac{\alpha_n}{4}, \frac{\alpha_n}{4}] & (i = n), \end{cases}$$

by the following formulas:

$$x_i = x_i(t_i) = \begin{cases} \int_{t_i}^{\beta_i} \frac{dt}{\sqrt{(-1)^{i-1} f_1(t) \ldots f_{n-1}(t)}} & (1 \le i \le n-1) \\ \int_{t_n}^{\beta_n} \frac{dt}{\sqrt{(-1)^{n-1} f_1(t) \ldots f_{n-1}(t)}} - \frac{\alpha_n}{4} & (i = n). \end{cases}$$

The inverse function $t_i(x_i)$ $(1 \le i \le n)$ can be uniquely extended as a C^∞ function

$$t_i : \boldsymbol{R}/\alpha_i \boldsymbol{Z} \to [\beta_{i-1}, \beta_i]$$

satisfying the differential equation

$$\left(\frac{dt_i}{dx_i}\right)^2 = (-1)^{i-1} f_1(t_i) \dots f_{n-1}(t_i).$$

We have $t_i(x_i) = t_i(-x_i)$ if $1 \le i \le n-1$ and $t_i(x_i) = t_i(\alpha_i/2 - x_i)$ if $2 \le i \le n$. Put

$$R = \prod_{i=1}^{n} (\boldsymbol{R}/\alpha_i \boldsymbol{Z}) = \{[x_1, \dots, x_n]\},$$

where $[x]$ denotes the image of $x = (x_1, \dots, x_n)$ by the natural projection $\boldsymbol{R}^n \to R$.

Let us define the involutions $\sigma_i : R \to R$ $(1 \le i \le n-1)$ by

$$\sigma_i([x]) = [x_1, \dots, x_{i-1}, -x_i, \frac{\alpha_{i+1}}{2} - x_{i+1}, x_{i+2}, \dots, x_n].$$

It is easily seen that σ_i and σ_j commute for any i, j. Let G be the group of transformations of R generated by $\sigma_1, \dots, \sigma_{n-1}$.

PROPOSITION 3.3.2. *The quotient space R/G possesses the natural differentiable structure, and it is diffeomorphic to S^n.*

PROOF. Let $\overline{I}_k$ be the set of fixed points of σ_k, i.e.,

$$\overline{I}_k = \{[x] \in T \mid x_k = 0 \text{ or } \frac{\alpha_k}{2},\ x_{k+1} = \frac{\alpha_{k+1}}{4} \text{ or } -\frac{\alpha_{k+1}}{4}\}.$$

Put $\overline{I} = \cup_{k=1}^{n-1} \overline{I}_k$. Note that $\overline{I}_k \cap \overline{I}_{k+1} = \emptyset$. Let $\Phi : R \to R/G$ be the quotient mapping, and put $\Phi(\overline{I}_k) = I_k$, $\Phi(\overline{I}) = I$. As is easily seen, G acts freely on $R - \overline{I}$. Hence the mapping

$$\Phi : R - \overline{I} \to R/G - I$$

is an ordinary (unbranched) 2^{n-1}-fold covering, and $R/G - I$ becomes a C^∞ manifold so that Φ is a local diffeomorphism.

Let $[a] \in \overline{I}$. Then there are $i_1, \dots, i_l$ such that $1 \le i_1 < \dots < i_l \le n-1$ and

$$[a] \in \cap_{m=1}^{l} \overline{I}_{i_m} - \cup_{j \ne i_1, \dots, i_l} \overline{I}_j. \tag{3.3.1}$$

Here we have $i_{m+1} - i_m \ge 2$. On a neighborhood of $[a]$ we define functions $y_1, \dots, y_n$ as follows:

$$\begin{aligned} y_{i_m} &= (x_{i_m} - a_{i_m})^2 - (x_{i_m+1} - a_{i_m+1})^2 \qquad (1 \le m \le l) \\ y_{i_m+1} &= 2(x_{i_m} - a_{i_m})(x_{i_m+1} - a_{i_m+1}) \qquad (1 \le m \le l) \\ y_j &= x_j \qquad (j \ne i_1, \dots, i_l, i_1+1, \dots, i_l+1). \end{aligned} \tag{3.3.2}$$

Clearly the system of functions $(y_1, \dots, y_n)$ is projectable, and gives a homeomorphism from a neighborhood of $\Phi([a])$ in R/G to an open subset of $\boldsymbol{R}^n$. It is also easily seen that those systems of functions for all $\Phi([a]) \in I$, together with C^∞ coordinate systems on $R/G - I$, form a C^∞ atlas of R/G. Thus R/G becomes a C^∞ manifold. We shall observe in Appendix that the manifold R/G so defined is actually diffeomorphic to the sphere S^n. This completes the proof. □

Next, we shall define functions $F_1, \dots, F_n$ on $T^*(R/G)$. Define functions $f_{ik} \in C^\infty(\boldsymbol{R}/\alpha_i\boldsymbol{Z})$ by

$$f_{ik}(x_i) = f_k(t_i(x_i)) \qquad (1 \le k \le n-1,\ 1 \le i \le n).$$

Let $B = (b_{ij})$ be the $n \times n$ matrix-valued function on R given by

$$b_{ij} = b_{ij}(x_i) = \begin{cases} (-1)^i \prod_{k \neq j} f_{ik}(x_i) & (1 \leq j \leq n-1) \\ (-1)^{i+1} \prod_k f_{ik}(x_i) & (j = n). \end{cases}$$

Let B_i^j denote the matrix of size $(n-1) \times (n-1)$ obtained by deleting the i-th row and the j-th column from B.

LEMMA 3.3.3. *On $R - \overline{I}$, we have:*
(1) $(-1)^{i+1} \det B_i^n > 0$;
(2) $(-1)^{n-1} \det B > 0$.

Let (x, ξ) be the canonical coordinate system associated with x. From the lemma above we can define $\overline{F}_1, \ldots, \overline{F}_n \in C^\infty(T^*(R - \overline{I}))$ by

$$B \begin{pmatrix} \overline{F}_1 \\ \vdots \\ \overline{F}_n \end{pmatrix} = \begin{pmatrix} \xi_1^2 \\ \vdots \\ \xi_n^2 \end{pmatrix}.$$

Clearly $\overline{F}_n$ is everywhere positive definite. Moreover, $\overline{F}_i$ being G-invariant, we can find $F_1, \ldots, F_n \in C^\infty(T^*(R/G - I))$ such that $\overline{F}_i \circ \Phi^* = F_i$. Lemma 3.3.3 follows from the next lemma.

LEMMA 3.3.4. *Let $A = (a_{ij})$ be a matrix of size $n \times (n-1)$, $n \geq 2$. Suppose that there are integers l $(0 \leq l \leq n-1)$ and i_k, j_k $(1 \leq k \leq l)$ with*

$$1 \leq i_1 < \cdots < i_l \leq n, \quad 1 \leq j_1 < \cdots < j_l \leq n-1$$

such that the following conditions are satisfied.
(1) *Each i_k is equal with j_k or $j_k + 1$. $a_{i_k j_k}$ is positive if $i_k = j_k$, and negative if $i_k = j_k + 1$.*
(2) $a_{i_k j} = 0$ *if $j \neq j_k$.*
(3) *When $\alpha \neq i_1, \ldots, i_l$, $a_{\alpha\beta}$ is negative if $\alpha \leq \beta$, and positive if $\alpha > \beta$.*
(4) *If $\alpha_1, \alpha_2 \neq i_1, \ldots, i_l$ and $1 \leq \alpha_1 \leq \beta_1 < \beta_2 < \alpha_2 \leq n$, then*

$$\frac{a_{\alpha_1\beta_1}}{a_{\alpha_1\beta_2}} > \frac{a_{\alpha_2\beta_1}}{a_{\alpha_2\beta_2}} \quad (> 0).$$

Then $(-1)^{i-1} \det A_i > 0$, where A_i denotes the matrix of size $(n-1) \times (n-1)$ obtained by deleting the i-th row from A.

PROOF. We shall only show the case where $l = 0$. Other cases will be proved almost in the same way. We shall prove this by induction on n. If $n = 2$, the assertion is clear. Let $n \geq 3$, and assume that the lemma is true for matrices of size $(n-1) \times (n-2)$. Let A_i^i denote the matrix of size $(n-1) \times (n-2)$ obtained by deleting the i-th row and the i-th column from A, and let A_{ik}^i denote the matrix of size $(n-2) \times (n-2)$ obtained by deleting the i-th and k-th rows and the i-th column from A. If $1 \leq i \leq n-1$, then A_i^i satisfies the conditions of the lemma as a matrix of size $(n-1) \times (n-2)$. Hence in this case we have

$$\begin{aligned} &(-1)^{k-1} \det A_{ik}^i > 0 \quad \text{and} \quad a_{ki} < 0 \quad \text{if} \quad 1 \leq k \leq i \\ &(-1)^{k-2} \det A_{ik}^i > 0 \quad \text{and} \quad a_{ki} > 0 \quad \text{if} \quad i < k \leq n. \end{aligned}$$

Expanding $\det A_i$ with respect to the i-th column, we have

$$(-1)^{i-1}\det A_i = \sum_{1\le k<i} (-1)^{k+i} a_{ki}(-1)^{k-1}\det A^i_{ik} + \sum_{i<k\le n} a_{ki}(-1)^k \det A^i_{ik}.$$

Note that each summand of the first sum in the right-hand side of the formula above is negative, and each summand of the second sum is positive. In particular we have $(-1)^{i-1}\det A_i > 0$ if $i = 1$.

Now, assume that $2 \le i \le n-1$. Then we have

$$\begin{aligned}(-1)^{i-1}\det A_i &> \max_{1\le k<i}\{\frac{a_{ki}}{a_{k,i-1}}\} \times \sum_{1\le k<i} a_{k,i-1}(-1)^{k-1}\det A^i_{ik} \\ &+ \min_{i<k\le n}\{\frac{a_{ki}}{a_{k,i-1}}\} \times \sum_{i<k\le n} a_{k,i-1}(-1)^k \det A^i_{ik}.\end{aligned}$$

Since

$$\min_{i<k\le n}\{\frac{a_{ki}}{a_{k,i-1}}\} > \max_{1\le k<i}\{\frac{a_{ki}}{a_{k,i-1}}\},$$

and since

$$\sum_{1\le k<i} a_{k,i-1}(-1)^{k-1}\det A^i_{ik} + \sum_{i<k\le n} a_{k,i-1}(-1)^k \det A^i_{ik} = 0,$$

it therefore follows that $(-1)^{i-1}\det A_i > 0$. In the case where $i = n$ we expand $\det A_i$ with respect to the $(n-1)$-th column. Then we obtain the same result. □

We now go back to the situation before Lemma 3.3.4. The following proposition will prove Theorem 3.3.1 in the case of type (A).

PROPOSITION 3.3.5. *$F_1, \dots, F_n$ are uniquely extended to $T^*(R/G)$ as C^∞ functions, and F_n is still positive definite there. Define a riemannian metric on R/G so that the associated energy function is $E = F_n/2$, and let $\mathcal{F}$ be the vector space spanned by $F_1, \dots, F_n$. Then $(R/G, \mathcal{F})$ is a proper Liouville manifold of rank one and type* (A)*, and its core is isomorphic to the one that first chosen.*

PROOF. First we show that F_i extends to every point $p \in I$. Let $[a] \in \bar{I}$ be as (3.3.1). Then, by the definition of $b_{ij}(x_i)$, one can easily see that if the Taylor expansion of $b_{i_m,j}$ $(1 \le m \le l)$ at $x_{i_m} = a_{i_m}$ is given by

$$b_{i_m,j}(x_{i_m}) \sim \sum_k \gamma_k (x_{i_m} - a_{i_m})^{2k},$$

then that of $b_{i_m+1,j}$ at $x_{i_m+1} = a_{i_m+1}$ becomes

$$b_{i_m+1,j}(x_{i_m+1}) \sim \sum_k (-1)^{k+1}\gamma_k (x_{i_m+1} - a_{i_m+1})^{2k}.$$

Hence it follows from Proposition 1.3.7 that F_i are smoothly defined around the point $\Phi([a])$ and satisfy the properness condition there. The property that F_n is positive definite at $\Phi([a])$ follows from the condition (A.2).

Put $I_m = \Phi(\bar{I}_m)$ $(1 \le m \le n-1)$. Then we have

$$I_m = \{q \in R/G \mid (F_m)_q = 0\} \qquad (1 \le m \le n-1).$$

It is also easily verified that the set

$$J_m = \{\Phi([x]) \mid x_m = 0 \text{ or } \frac{\alpha_m}{2}, \text{ or } x_{m+1} = \frac{\alpha_{m+1}}{4} \text{ or } -\frac{\alpha_{m+1}}{4}\}$$

is identical with the set

$$\{q \in R/G \mid \operatorname{rank}(F_m)_q \leq 1\}.$$

Therefore it is now clear that $(R/G, \mathcal{F})$ is a proper Liouville manifold of rank one and type (A), and the core is isomorphic with the one that we first took. □

Next, let us observe the other cases. The case X = B is reduced to the case X = A by taking the double covering. Now let X = C. In this case we construct

$$R = \prod_{i=1}^{n-1} (\boldsymbol{R}/\alpha_i \boldsymbol{Z}) \times (-\alpha_n, \alpha_n),$$

where $0 < \alpha_n \leq \infty$. For $i = 1, \ldots, n-1$, α_i and the coordinate functions $x_i = x_i(t_i)$ are given by the same formulas as in the case X = A. We put

$$x_n = x_n(t_n) = \int_{\beta_{n-1}}^{t_n} \frac{dt}{\sqrt{(-1)^{n-1} f_1(t) \ldots f_{n-1}(t)}} \qquad (\beta_{n-1} \leq t_n < \infty)$$

and $x_n(\infty) = \alpha_n$. Then as before the inverse function $t_n(x_n)$ can be uniquely extended as the C^∞ function

$$t_n : (-\alpha_n, \alpha_n) \to [\beta_{n-1}, \infty)$$

so that $t_n(-x_n) = t_n(x_n)$. We also define the group G of transformations of R in the same way except σ_{n-1}, which is now given by

$$\sigma_{n-1}([x]) = [x_1, \ldots, x_{n-2}, -x_{n-1}, -x_n].$$

Then the quotient space R/G becomes a C^∞ manifold, and in view of Appendix it is diffeomorphic to $\boldsymbol{R}^n$. The remaining part of the construction is the same as in the case of type (A). The completeness of the resulting riemannian manifold R/G is also easily verified.

Finally, we consider the case where X = D. In this case the variables x_1 and x_n are given by the different formulas from those in the case X = A:

$$x_1 = x_1(t_1) = \int_{t_1}^{\beta_1} \frac{dt}{\sqrt{f_1(t) \ldots f_{n-1}(t)}} \qquad (-\infty < t_1 \leq \beta_1),$$

$$x_n = x_n(t_n) = \int_{\beta_{n-1}}^{t_n} \frac{dt}{\sqrt{(-1)^{n-1} f_1(t) \ldots f_{n-1}(t)}} \qquad (\beta_{n-1} \leq t_n < \infty).$$

We put $x_1(-\infty) = \alpha_1$ and $x_n(\infty) = \alpha_n$. Note that these numbers may be ∞. The inverse functions $t_1(x_1)$ and $t_n(x_n)$ are uniquely extended to $(-\alpha_1, \alpha_1)$ and $(-\alpha_n, \alpha_n)$ respectively as C^∞ even functions. We put

$$R = (-\alpha_1, \alpha_1) \times \prod_{i=2}^{n-1} (\boldsymbol{R}/\alpha_i \boldsymbol{Z}) \times (-\alpha_n, \alpha_n) = \{[x_1, \ldots, x_n]\}$$

Correspondingly, the definitions of σ_1 and σ_{n-1} in this case are given as follows:

$$\sigma_1([x]) = [-x_1, \alpha_2/2 - x_2, x_3, \ldots, x_n],$$
$$\sigma_{n-1}([x]) = [x_1, \ldots, x_{n-2}, -x_{n-1}, -x_n].$$

Then the quotient space R/G becomes a C^∞ manifold, and we can see that it is diffeomorphic to $\boldsymbol{R}^n$ by Appendix. The remaining part is the same as in the case X = A. This finishes the proof of Theorem 3.3.1.

3.4. Classification

In this section we shall prove the following

THEOREM 3.4.1. *The assignment of the core to a proper Liouville manifold gives the one-to-one correspondence between the isomorphism classes of proper Liouville manifolds of rank one and the isomorphism classes of possible cores.*

REMARK. This theorem combined with the results in the previous section indicates that any proper Liouville manifold of rank one is isomorphic to the one that constructed from the core. In particular it is diffeomorphic to S^n or $\boldsymbol{RP}^n$ or $\boldsymbol{R}^n$ according to the type.

We first assume that the type of $(M, \mathcal{F})$ is (A) or (C) or (D) so that M is simply connected. By virtue of Theorem 3.3.1, to prove the theorem in these cases it suffices to show the next proposition .

PROPOSITION 3.4.2. *Let $(\widetilde{M}, \widetilde{\mathcal{F}})$ and $(M, \mathcal{F})$ be n-dimensional proper Liouville manifolds of rank one and type (X) $(X \neq B)$. Assume that there is an isomorphism $\psi : \cap_k \widetilde{J}_k \to \cap_k J_k$ of the cores. Let*

$$\psi_1 : T\widetilde{M}|_{\cap_k \widetilde{J}_k} \to TM|_{\cap_k J_k}$$

be a bundle isomorphism such that:
(1) *ψ_1 induces ψ;*
(2) *ψ_1 is fibrewise isometric;*
(3) *ψ_1 maps $N\widetilde{J}_m|_{\cap_k \widetilde{J}_k}$ to $NJ_m|_{\cap_k J_k}$ $(1 \leq m \leq n-1)$.*
Then there is a unique isomorphism $\Psi : (\widetilde{M}, \widetilde{\mathcal{F}}) \to (M, \mathcal{F})$ of Liouville manifolds such that the restriction of Ψ_* to $T\widetilde{M}|_{\cap_k \widetilde{J}_k}$ coincides with ψ_1. Here,

$$\widetilde{J}_k = \{q \in \widetilde{M} \mid \mathrm{rank}\,(\widetilde{F}_k)_q \leq 1\}$$

for representatives $\widetilde{F}_k$ of $\widetilde{\mathcal{F}}^*/\sim$, and their ordering is given as Proposition 3.1.3.

We shall prove this proposition by induction on $n = \dim \widetilde{M} = \dim M$. The case $n = 2$ is an immediate consequence of Theorem 3.1 in **[Ki1]**, Theorem 3.1 in **[IKS]**, and Theorem 3.3.1. So, let $n \geq 3$ and assume that the proposition is true for manifolds of dimension less than n. We first consider the case of type (A). Without loss of generality, we may assume that the Liouville manifold $(\widetilde{M}, \widetilde{\mathcal{F}})$ is the one constructed from its core. By the assumption, we can choose representatives $\widetilde{F}_m$ and F_m so that $\psi_1(\widetilde{F}_m) = F_m$ $(1 \leq m \leq n-1)$. Also, we may assume that both core submanifolds are parametrized by $\boldsymbol{R}/l\boldsymbol{Z}$ as in Proposition 3.1.3 so that ψ is expressed by the identity of $\boldsymbol{R}/l\boldsymbol{Z}$.

In order to complete the induction we first observe some properties of admissible frames and admissible submanifolds of $(M, \mathcal{F})$. Since M is simply connected, there are global sections $V_1^2, \dots, V_n^2$ of $\mathcal{S}^2TM^0$ such that at each point in M^0, their square roots $V_1, \dots, V_n$ form an admissible basis. We shall take those as follows: Let p_0 be the midpoint of the interval $(-\beta_1, \beta_1)$ in $\cap_m J_m = \boldsymbol{R}/l\boldsymbol{Z}$, and take the admissible basis $V_1, \dots, V_n$ of $T_{p_0}M$ so that

$$F_m = f_m V_{m+1}^2, \; f_m > 0 \quad (1 \le m \le n-1), \quad V_1 = \frac{\partial}{\partial t}$$

at p_0. Then, extend V_m^2 to the whole M^0 (uniquely).

LEMMA 3.4.2.

$$V_m^2 = (\partial/\partial t)^2 \quad on \quad \begin{cases} \cup_{i,j=0}^1 \sigma^i \tau^j ((\beta_{m-1}, \beta_m)) & if \;\; 2 \le m \le n-1 \\ \cup_{j=0}^1 \tau^j ((-\beta_1, \beta_1)) & if \;\; m = 1 \\ \cup_{i=0}^1 \sigma^i ((\beta_{n-1}, l/2 - \beta_{n-1})) & if \;\; m = n, \end{cases}$$

where σ and τ are the diffeomorphism of $\boldsymbol{R}/l\boldsymbol{Z}$ given by $\sigma(t) = -t$ and $\tau(t) = -t + l/2$.

PROOF. Note that V_{m+1}^2 extends to the connected component of $J_m - I_m$ containing p_0 so that its square root is normal to J_m, and spans a simultaneous eigenspace of $\{F^e \mid F \in \mathcal{F}\}$. Therefore, on the interval $(-\beta_m, \beta_m)$ in $\cap_k J_k$, V_{m+1} is normal to J_m. If $V_i^2 = (\partial/\partial t)^2$ on the interval (β_{m-1}, β_m), then the admissible subspace $N_{\beta_m} I_m$ is connected to an admissible subbundle on M^0 spanned by V_{m+1} and V_i. This implies $V_{m+1}^2 = (\partial/\partial t)^2$ on the interval (β_m, β_{m+1}). Similarly, we have the same result on the interval $(-\beta_{m+1}, -\beta_m)$.

The results above indicate that on the intervals (β_{n-1}, β_n) and $(-\beta_n, -\beta_{n-1})$

$$V_n^2 = \left(\frac{\partial}{\partial t}\right)^2, \quad F_m = f_m V_m^2.$$

Hence in a similar way as above, we have the equalities on the remaining intervals.□

LEMMA 3.4.3. (1) $I_{m-1} \cap I_m = \emptyset$.

(2) *V_m^2 is extendable to $M - I_{m-1} \cup I_m$ for any m $(1 \le m \le n)$, where $I_0 = I_n = \emptyset$.*

PROOF. It is easily seen from the previous lemma that the normal bundle of I_m is smoothly connected to the subbundle of TM^0 spanned by V_m and V_{m+1}. This implies that $I_{m-1} \cap I_m = \emptyset$, and V_m is smoothly extendable to a point $p \in M^s$ if and only if $p \notin I_{m-1}$ and $p \notin I_m$. □

Let p_t (resp. $\tilde{p}_t$) be the point on $\cap_k J_k \cap M^0$ (resp. on $\cap_k \tilde{J}_k \cap \widetilde{M}^0$) that is represented by $t \in \boldsymbol{R}/l\boldsymbol{Z}$. Let $W \subset T_{p_t}M$ be the normal subspace to the core submanifold, and let $N(p_t)$ be the maximal admissible submanifold determined by W.

PROPOSITION 3.4.4. *Suppose $t \in (-\beta_1, \beta_1)$ (resp. $t \in (\beta_{n-1}, l/2 - \beta_{n-1})$). Then $N = N(p_t)$ possesses the following properties:*

(1) *The proper Liouville manifold N is of rank one and type* (A), *and its core submanifold is given by $\cap_{k=2}^{n-1} J_k \cap N$ (resp. $\cap_{k=1}^{n-2} J_k \cap N$);*
(2) *the set $\cap_k J_k \cap N$ consists of the two points represented by t and $l/2 - t$.*

PROOF. Suppose first that $t \in (-\beta_1, \beta_1)$. Let F'_m and E' be the pull-back images of F_m and E by the embedding $T^*N \to T^*M$ respectively, and let $\mathcal{F}'$ be the vector space spanned by them. Then, as observed in Proposition 2.2.3, $(N, \mathcal{F}')$ is a proper Liouville manifold and $N^s = N \cap M^s$. We consider the 2-dimensional Liouville manifold $\cap_{k\geq 2} J_k$. The intersection $N \cap \cap_{k\geq 2} J_k$ contains the coordinate line through p, which intersects every I_m $(m \geq 2)$ orthogonally. Hence on a neighborhood of the intersection points, $I'_m = I_m \cap N$ is a submanifold of N of codimension 2. This implies that $F'_m \in (\mathcal{F}')^*$ for any $m \geq 2$, and it follows that N is of rank one. Put

$$J'_m = \{q \in N \mid \operatorname{rank}(F'_m)_q \leq 1\}.$$

Then we immediately have

$$\cap_{m\geq 2} J'_m \quad \supset \quad \cap_{m\geq 2} J_m \cap N \quad \supset \quad \text{the coordinate line through } p.$$

Among them, the left one is a connected, 1-dimensional manifold. So being the right one, we see that all of them actually coincide. This shows (2). Now let us consider the intersection of I_m with $\cap_{k\geq 2} J'_k$. Since $I_m \cap \cap_{k\geq 2} J_k$ consists of the two coordinate lines through β_m and $l/2 - \beta_m$ respectively, it follows that

$$I_m \cap \cap_{k\geq 2} J'_k = I_m \cap \cap_{k\geq 2} J_k \cap N$$

consists of 4 points. Hence N is of type (A), and we have (1). The case where $t \in (\beta_{n-1}, l/2 - \beta_1)$ is similar. □

Let W_k (resp. $\widetilde{W}_k$) be a unit normal vector field to J_k (resp. to $\widetilde{J}_k$) $(1 \leq k \leq n-1)$ so that $\psi(\widetilde{W}_k) = W_k$ on $\cap_k \widetilde{J}_k$. We now extend ψ and ψ_1 to $\cup_k \widetilde{J}_k$.

LEMMA 3.4.5. *ψ and ψ_1 are uniquely extended to $\cup_k \widetilde{J}_k$ and $T\widetilde{M}|_{\cup_k \widetilde{J}_k}$ respectively so that they satisfy the following conditions:*

(1) $\psi_1 : T\widetilde{M}|_{\widetilde{J}_k} \to TM|_{J_k}$ is a bundle isomorphism for each k;

(2) ψ_1 induces ψ;

(3) $\psi_1(\widetilde{W}_k) = W_k$ on J_k;

(4) $\psi : \widetilde{J}_k \to J_k$ are isomorphisms of Liouville manifolds;

(5) $\psi_1 = (\psi|_{\widetilde{J}_l})_*$ on $N\widetilde{J}_k|_{\widetilde{J}_k \cap \widetilde{J}_l}$ for any k and l $(\neq k)$.

Moreover, the extended ψ_1 satisfies $\psi_1(\widetilde{F}_m) = F_m$ for every m.

PROOF. Note that $\widetilde{J}_k$ (resp. J_k) is a proper Liouville manifold of rank one and type (A), and the core submanifold is the same one as that of $(\widetilde{M}, \widetilde{\mathcal{F}})$ (resp. $(M, \mathcal{F})$). Also, since $\psi_1(\widetilde{F}_m) = F_m$, the mapping $\psi : \cap_m \widetilde{J}_m \to \cap_m J_m$ gives an isomorphism between the cores of $\widetilde{J}_k$ and J_k. Hence by induction assumption, ψ is uniquely extended to an isomorphism $\psi : \widetilde{J}_k \to J_k$ of Liouville manifolds so that $\psi_* = \psi_1$ on $\cap \widetilde{J}_m$. In particular, we have $\psi_*(\widetilde{W}_m) = W_m$ $(m \neq k)$ on $\widetilde{J}_k \cap \widetilde{J}_m$ and $\psi_*(\widetilde{F}'_m) = F'_m$, where $\widetilde{F}'_m$ (resp. F'_m) denotes restriction of $\widetilde{F}_m$ to $T^*\widetilde{J}_k$ (resp. F_m to T^*J_k). We define $\psi_1 : T\widetilde{M}|_{\widetilde{J}_k} \to TM|_{J_k}$ by $\psi_1 = \psi_*$ on $T\widetilde{J}_k$ and $\psi_1(\widetilde{W}_k) = W_k$. Then, they clearly satisfy the conditions above. The uniqueness is also obvious.

We now prove the last assertion. Let $\widetilde{V}_m$ $(1 \leq m \leq n)$ be the admissible orthonormal frame of $T\widetilde{M}^0$ defined in the same way as V_m. Let $\widetilde{J}_k^+$ (resp. $\widetilde{J}_k^-$)

be the subset of $\widetilde{J}_k$ where $\widetilde{F}_k$ is positive (resp. negative) semi-definite and non-zero. As is easily seen, $\widetilde{V}_{k+1}$ (resp. $\widetilde{V}_k$) is normal to $\widetilde{J}_k^+$ (resp. to $\widetilde{J}_k^-$). We define $J_k^{\pm}$ similarly. Clearly we have $\psi(\widetilde{J}_k^+) = J_k^+$ and $\psi(\widetilde{J}_k^-) = J_k^-$. (Note that the connected components of $\widetilde{I}_k$ and I_k are the maximal admissible submanifolds of $\widetilde{J}_k$ and J_k respectively, so $\psi(\widetilde{I}_k) = I_k$.)

We first consider ψ and ψ_1 on $\widetilde{J}_k^-$. Describing $\widetilde{V}_i^2 = \sum_j \widetilde{a}_{ij}\widetilde{F}_j$ and $V_i^2 = \sum_j a_{ij}F_j$ respectively, we have $0 = \widetilde{a}_{kj}$ $(j \neq k)$ and $\widetilde{V}_i^2 = \sum_{j\neq k} \widetilde{a}_{ij}\widetilde{F}_j'$ $(i \neq k)$ on $\widetilde{J}_k^-$, and the same formula for F_j' on J_k^-. Since $\psi_*(\widetilde{V}_j^2) = V_j^2$ and $\psi_*(\widetilde{F}_j') = F_j'$ $(j \neq k)$, it follows that $\psi^* a_{ij} = \widetilde{a}_{ij}$ for $i, j \neq k$. Let $\widetilde{q} \in \widetilde{J}_k^- \cap \widetilde{J}_k^0$ and let $\widetilde{N}$ be the maximal admissible submanifold of $\widetilde{J}_k$ determined by the normal subspace to V_l $(l \neq k)$ in $T_{\widetilde{q}}\widetilde{J}_k$. Since $\widetilde{M}$ (and hence $\widetilde{J}_k$) is assumed to be the one that constructed from the core, it easily follows from the previous section that $\widetilde{N}$ intersects the interval (β_{l-1}, β_l) in $\cap_m \widetilde{J}_m$ exactly at one point, say $\widetilde{p}_t$. Put $\psi(\widetilde{N}) = N$ and $\psi(\widetilde{q}) = q$. Since the ratio

$$(\widetilde{a}_{l1} : \cdots : \widetilde{a}_{ln}) \in \boldsymbol{R}P^{n-1}$$

is constant along $\widetilde{N}$ (cf. Proposition 1.1.3), and so is on N, we have

$$\begin{aligned}(\widetilde{a}_{l1}(\widetilde{q}) : \cdots : \widetilde{a}_{ln}(\widetilde{q})) &= (\widetilde{a}_{l1}(\widetilde{p}_t) : \cdots : \widetilde{a}_{ln}(\widetilde{p}_t)) \\ &= (a_{l1}(p_t) : \cdots : a_{ln}(p_t)) = (a_{l1}(q) : \cdots : a_{ln}(q)).\end{aligned}$$

Since $\widetilde{a}_{lm}(\widetilde{q}) = a_{lm}(q)$ for $m \neq k$, it therefore follows that $\widetilde{a}_{lk}(\widetilde{q}) = a_{lk}(q)$ $(l \neq k)$. Since $\sum_{l=1}^n \widetilde{a}_{lk} = \sum_{l=1}^n a_{lk} = 0$, we also have $\widetilde{a}_{kk}(\widetilde{q}) = a_{kk}(q)$. Therefore we have $\psi^*(a_{ij}) = \widetilde{a}_{ij}$ for any i, j. Since $\psi_1(\widetilde{V}_i^2) = V_i^2$, we consequently obtain $\psi_1(\widetilde{F}_m) = F_m$ on $\widetilde{J}_k^- \cap J_k^0$ for any m. We have the same result on $\widetilde{J}_k^+ \cap \widetilde{J}_k^0$ by a similar argument. Hence we have $\psi_1(\widetilde{F}_m) = F_m$ on the whole $\widetilde{J}_k$ by continuity. □

LEMMA 3.4.6. *Let $t \in (-\beta_1, \beta_1) \cup (\beta_{n-1}, l/2 - \beta_{n-1})$. Then there are unique mappings $\Psi : N(\widetilde{p}_t) \to N(p_t)$ and $\Psi_1 : T\widetilde{M}|_{N(\widetilde{p}_t)} \to TM|_{N(p_t)}$ satisfying the following conditions:*

(1) *$\Psi : N(\widetilde{p}_t) \to N(p_t)$ is an isomorphism of Liouville manifolds;*
(2) *$\Psi_1 : T\widetilde{M}|_{N(\widetilde{p}_t)} \to TM|_{N(p_t)}$ is a bundle isomorphism preserving the fibre metrics;*
(3) *$\Psi_1 = \Psi_*$ on $TN(\widetilde{p}_t)$;*
(4) *$\Psi_1 = \psi_1$ on $T\widetilde{M}|_{\cup_k \widetilde{J}_k \cap N(\widetilde{p}_t)}$.*

Moreover, Ψ_1 satisfies $\Psi_1(\widetilde{F}_m) = F_m$ for any m, and $\Psi : N(\widetilde{p}_t) \to N(p_t)$ $(t \in (-\beta_1, \beta_1))$ and $\Psi : N(\widetilde{p}_s) \to N(p_s)$ $(s \in (\beta_{n-1}, l/2 - \beta_{n-1}))$ coincide on $N(\widetilde{p}_t) \cap N(\widetilde{p}_s)$.

PROOF. We first assume that $t \in (-\beta_1, \beta_1)$. From the previous lemma we have the isomorphism $\psi : \cap_{m\geq 2}\widetilde{J}_m \to \cap_{m\geq 2}J_m$ of Liouville surfaces and the bundle isomorphism $\psi_1 : T\widetilde{M}|_{\cap_{m\geq 2}\widetilde{J}_m} \to TM|_{\cap_{m\geq 2}J_m}$ so that $\psi_1(\widetilde{F}_k) = F_k$ for any k. Hence by Proposition 3.4.4 and the induction assumption, one obtains a unique isomorphism $\Psi : N(\widetilde{p}_t) \to N(p_t)$ of Liouville manifolds so that $\Psi_* = \psi_1$ on $TN(\widetilde{p}_t)|_{\cap_{m\geq 2}\widetilde{J}_m \cap N(\widetilde{p}_t)}$. We define $\Psi_1 : T\widetilde{M}|_{N(\widetilde{p}_t)} \to TM|_{N(p_t)}$ by $\Psi_1 = \Psi_*$ on $TN(\widetilde{p}_t)$ and $\Psi(\widetilde{V}_1) = V_1$. Here, we took $\widetilde{V}_1$, the square root of $\widetilde{V}_1^2$, so that it is smooth along $N(\widetilde{p}_t)$ and it coincides with $\partial/\partial t$ at $\widetilde{p}_t \in (-\beta_1, \beta_1)$, and so did for V_1. It is possible, because the normal bundle of $N(\widetilde{p}_t)$ is trivial.

The property $\Psi_1(\widetilde{F}_m) = F_m$ is verified in the same way as the proof of the previous lemma. To prove (4) it suffices to show $\Psi = \psi$ on $\widetilde{J}'_k = N(\widetilde{p}_t) \cap \widetilde{J}_k$ for every k. Clearly we have $\Psi(\widetilde{J}'_k) = \psi(\widetilde{J}'_k) = J'_k$, where $J'_k = N(p_t) \cap J_k$. In case $n = 3$, these submanifolds are of dimension 1. Since $\Psi_* = \psi_*$ on $T_{\widetilde{p}_t}\widetilde{J}'_k$ and Ψ and ψ are isometries from $\widetilde{J}'_k$ to J'_k, we have $\Psi = \psi$ on $\widetilde{J}'_k$.

Now, let $n \geq 4$ and assume that $k \geq 2$. Then the submanifolds $\widetilde{J}'_k$ and J'_k are proper Liouville manifold of rank one and type (A). The core submanifold $\widetilde{C}$ (resp. C) of $\widetilde{J}'_k$ (resp. J'_k) is equal to $N(\widetilde{p}_t) \cap \cap_{m\geq 2} \widetilde{J}_m$ (resp. $N(p_t) \cap \cap_{m\geq 2} J_m$). From the construction of $\Psi : N(\widetilde{p}_t) \to N(p_t)$ and $\psi : \widetilde{J}_k \to J_k$ it follows that $\Psi_* = \psi_*$ on $T\widetilde{J}'_k|_{\widetilde{C}}$. Since the extended isomorphism $\widetilde{J}'_k \to J'_k$ should be unique by the induction assumption, we have $\Psi = \psi$ on $\widetilde{J}'_k$.

Next we consider $\widetilde{J}'_1$. As is easily seen from the previous section, the core submanifold $\widetilde{C}$ (resp. C) of $\widetilde{J}'_1$ (resp. J'_1) is equal to $N(\widetilde{p}_t) \cap \cap_{m\neq 2} \widetilde{J}_m$ (resp. $N(p_t) \cap \cap_{m\neq 2} J_m$). Then, by the fact that we have just proved we get $\Psi_* = \psi_*$ on $T\widetilde{J}'_1|_{\widetilde{C}}$. Thus by the same reason as above, we have $\Psi = \psi$ on $\widetilde{J}'_1$.

The last assertion is proved as follows. Let q be one of the intersection points of two coordinate lines $N(p_t) \cap \cap_{m\geq 2} J_m$ and $N(p_s) \cap \cap_{m\geq 2} J_m$ on $\cap_{m\geq 2} J_m$. Let K be the connected component of $N(p_t) \cap N(p_s)$ through q. Then K is the maximal admissible submanifold of $N(p_t)$ determined by the normal subspace to the core submanifold at q. Suppose that there is another connected component K' of $N(p_t) \cap N(p_s)$. Clearly K' is also a maximal admissible submanifold of $N(p_t)$. Since $N(p_t)$ is isomorphic to the constructed one, we see, using the branched covering Φ in Section 3.3, that K' also intersects the core submanifold of $N(p_t)$ at a point q' belonging to the same interval of $N(p_t) \cap \cap_{m\geq 2} J_m \cap M^0$ as q. Then the coordinate line on $\cap_{m\geq 2} J_m$ through q' is contained in $N(p_s)$ and passes through a point on the interval $(\beta_{n-1}, l/2 - \beta_{n-1})$ different from p_s. But this contradicts Proposition 3.4.4 (2). Hence $N(p_t) \cap N(p_s)$ is connected. By Proposition 3.4.4 it is a proper Liouville manifold of rank one and type (A) whose core submanifold is

$$N(p_t) \cap N(p_s) \cap_{m=2}^{n-2} J_m.$$

Hence the last assertion follows from the uniqueness in the induction assumption.□

We now define the mapping $\Psi : \widetilde{M} \to M$ so that $\Psi = \psi$ on $\cup_k \widetilde{J}_k$ and $\Psi|_{N(\widetilde{p}_t)}$ is the given one for $t \in (-\beta_1, \beta_1)$ and for $t \in (\beta_{n-1}, l/2 - \beta_{n-1})$. From Section 3.3 one can easily see that $\widetilde{M} = \widetilde{J}_1 \cup \cup_{|t|<\beta_1} N(\widetilde{p}_t)$. Hence, by virtue of Lemmas 3.4.5 and 3.4.6 the mapping $\Psi : \widetilde{M} \to M$ is well-defined and injective. To prove that Ψ is an isomorphism of Liouville manifolds, we need the following lemmas.

LEMMA 3.4.7. *The mapping Ψ is smooth on $\widetilde{M} - \widetilde{I}_1$, and $\Psi_* = \Psi_1$ on $T(\widetilde{M} - \widetilde{I}_1)$.*

PROOF. Let $\widetilde{D}_1$ (resp. D_1) be the subbundle of $T(\widetilde{M} - \widetilde{I}_1)$ (resp. $T(M - I_1)$) spanned by $\widetilde{V}_1$ (resp. V_1). Since $\widetilde{V}_1^2$ and V_1^2 are defined on $\widetilde{M} - \widetilde{I}_1$ and $M - I_1$ respectively, $\widetilde{D}_1$ and D_1 are well-defined. Let $\widetilde{D}_1^\perp$ and $D_1^\perp$ be the orthogonal complement of $\widetilde{D}_1$ and D_1 respectively. Note that $\widetilde{N}(\widetilde{p}_t)$ ($|t| < \beta_1$) and $\widetilde{J}_1^-$ (resp. $N(p_t)$ and J_1^-) are integral manifolds of $\widetilde{D}_1^\perp$ (resp. of $D_1^\perp$).

Let $\widetilde{q}_0 \in \widetilde{M} - \widetilde{I}_1$ be an arbitrary point, and put $\Psi(\widetilde{q}_0) = q_0$. Let $\widetilde{U}'$ be a small integral manifold of $\widetilde{D}_1^\perp$ of dimension $n-1$ through $\widetilde{q}_0$, and put $\Psi(\widetilde{U}') = U'$. By Lemmas 3.4.5 and 3.4.6, U' is an integral manifold of $D_1^\perp$, and $\Psi : \widetilde{U}' \to U'$ is a diffeomorphism. According to the product structure $\widetilde{D}_1 + \widetilde{D}_1^\perp$, we have a diffeomorphism $\widetilde{c} = \widetilde{c}(r, \widetilde{q})$ from $(-\epsilon, \epsilon) \times \widetilde{U}'$ to a neighborhood $\widetilde{U}$ of $\widetilde{q}_0$ so that $\widetilde{c}_*(\partial/\partial r) \in \widetilde{D}_1$ and $\widetilde{c}_*(T\widetilde{U}') = \widetilde{D}_1^\perp$ at each point. Similarly, we have a diffeomorphism $c = c(s, q)$ from $(-\delta, \delta) \times \widetilde{U}'$ to a neighborhood U of q_0.

Since $\Psi : \widetilde{J}_k \to J_k$ $(k \geq 2)$ and $\Psi : N(\widetilde{p}_t) \to N(p_t)$ $(t \in (\beta_{n-1}, l/2 - \beta_{n-1})$ are isomorphisms of Liouville manifolds, it follows that integral curves of $\widetilde{D}_1$ are mapped by Ψ to integral curves of D_1. Also, integral manifolds of $\widetilde{D}_1^\perp$ are mapped by Ψ to integral manifolds of $D_1^\perp$. Hence we have

$$c^{-1} \circ \Psi \circ \widetilde{c}(r, q) = (s(r), \Psi(q)), \quad q \in \widetilde{U}',$$

where $s(r)$ is a smooth function. This indicates that the mapping Ψ is smooth on $\widetilde{M} - \widetilde{I}_1$ and $\Psi_* = \Psi_1$ on $T(\widetilde{M} - \widetilde{I}_1)$. □

LEMMA 3.4.8. *The mapping $\Psi : \widetilde{M} \to M$ is smooth on a neighborhood of each point of $\widetilde{I}_1$.*

PROOF. Let $\widetilde{q}_0 \in \widetilde{I}_1$ be an arbitrary point and put $q_0 = \Psi(\widetilde{q}_0) \in I_1$. Let $\widetilde{U}'$ be a small neighborhood of $\widetilde{q}_0$ in $\widetilde{I}_1$, and put $U' = \Psi(\widetilde{U}')$. Since $\widetilde{U}' \subset \widetilde{J}_1$, it follows that U' is an neighborhood of q_0 in I_1 and $\Psi : \widetilde{U}' \to U'$ is a diffeomorphism. Let $\widetilde{D}_{12}$ be the subbundle of $T\widetilde{M}$ defined on a neighborhood of $\widetilde{q}_0$ spanned by $\widetilde{V}_1$ and $\widetilde{V}_2$ outside $\widetilde{I}_1$, and equal to the normal space $N\widetilde{I}_1$ on $\widetilde{I}_1$. D_{12} is defined similarly. Let $\widetilde{D}_{12}^\perp$ and $D_{12}^\perp$ be their orthogonal complements.

As observed in Section 1.1, these subbundles gives a local product structure. Hence there is a neighborhood $\widetilde{U}$ of $\widetilde{q}_0$ in $\widetilde{M}$ that is identified with $\widetilde{U}' \times B(\widetilde{q}_0)$, where $B(\widetilde{q}_0)$ is an integral manifold of $\widetilde{D}_{12}$ through $\widetilde{q}_0$. Similarly, there is a neighborhood U of q_0 that is identified with $U' \times B(q_0)$. We note that integral manifolds of $\widetilde{D}_{12}$ are contained in $\widetilde{J}_n^-$ or $N(\widetilde{p}_t)$ $(t \in (\beta_{n-1}, l/2 - \beta_{n-1}))$. Similarly, integral manifolds of $\widetilde{D}_{12}^\perp$ are contained in $\widetilde{J}_1$ or $N(\widetilde{p}_t)$ $(t \in (-\beta_1, \beta_1))$. Therefore, by virtue of Lemmas 3.4.5 and 3.4.6, $\Psi|_{\widetilde{U}}$ is represented by

$$\Psi : \widetilde{U}' \times B(\widetilde{q}_0) \to U' \times B(q_0), \quad \Psi(\widetilde{q}, \widetilde{p}) = (\Psi(\widetilde{q}), \Psi(\widetilde{p})).$$

This formula indicates that $\Psi : \widetilde{U} \to U$ is a diffeomorphism. □

We now prove Proposition 3.4.2 for Liouville manifolds of type (A). By Lemmas 3.4.7 and and 3.4.8 we see that the mapping Ψ is a diffeomorphism from $\widetilde{M}$ to an open subset of M. Also, by Lemma 3.4.7 we have $\Psi_*(\widetilde{F}_m) = F_m$ $(1 \leq m \leq n-1)$ and $\Psi_*(\widetilde{E}) = E$, where $\widetilde{E}$ is the energy function of $\widetilde{M}$. In particular, Ψ is an isometry. Since both $\widetilde{M}$ and M are complete, it follows that $\Psi(\widetilde{M}) = M$. Thus $\Psi : \widetilde{M} \to M$ is an isomorphism of Liouville manifolds. The uniqueness of Ψ is clear from the proof.

Next we shall consider the case of type (C) and (D). The proof of Proposition 3.4.2 for these cases are almost the same as for the case of type (A). The only difference is the type of the admissible submanifolds $N(p_t)$, which we now explain. Let $(M, \mathcal{F})$ be a proper Liouville manifold of rank one and type (C). Identifying the core submanifold $\cap_m J_m$ with $\boldsymbol{R}$ as Proposition 3.1.3, we have the following

LEMMA 3.4.9. *The admissible submanifold $N(p_t)$ is of rank one if $|t| < \beta_1$ or $|t| > \beta_{n-1}$. It is of type* (C) *if $|t| < \beta_1$ and of type* (A) *if $|t| > \beta_{n-1}$.*

For a Liouville manifold of type (D) we have the following

LEMMA 3.4.10. *The admissible submanifold $N(p_t)$ is of rank one and type* (C) *if $t < \beta_1$ or $t > \beta_{n-1}$.*

The proofs are similar to that of Lemma 3.4.4. Let $(\widetilde{M}, \widetilde{\mathcal{F}})$ be the Liouville manifold constructed from the core of $(M, \mathcal{F})$, and let $\psi_1 : T\widetilde{M}|_{\cap_m \widetilde{J}_m} \to TM|_{\cap_m J_m}$ be a bundle isomorphism satisfying the conditions in Proposition 3.4.2. Then we have $\widetilde{M} = \widetilde{J}_1 \cup \bigcup_{|t|<\beta_1} N(\widetilde{p}_t)$ for the case of type (C) and $\widetilde{M} = \widetilde{J}_1 \cup \bigcup_{t<\beta_1} N(\widetilde{p}_t)$ for the case of type (D).

Hence, as the case of type (A), we obtain the isomorphisms $\widetilde{J}_k \to J_k$ and $N(\widetilde{p}_t) \to N(p_t)$ by the induction assumption, and collecting them, we have the isomorphism $\Psi : \widetilde{M} \to M$ of Liouville manifolds. This completes the proof of Proposition 3.4.2. □

Finally, we shall give a proof of Theorem 3.4.1 for Liouville manifolds of type (B). Let $(M_i, \mathcal{F}_i)$ $(i = 1, 2)$ be two proper Liouville manifolds of rank one and type (B) with mutually isomorphic cores. Let $(\overline{M}_i, \overline{\mathcal{F}}_i)$ be their universal (double) coverings, and let τ_i be the covering transformations. Since the covering manifolds are of type (A), and since their cores are mutually isomorphic, we have an isomorphism

$$\Psi : \overline{M}_1 \to \overline{M}_2$$

of Liouville manifolds by Proposition 3.4.2.

Moreover, since τ_i induces the antipodal mapping on the core submanifold, and τ_{i*} reverses the orientation of NJ_m, we have

$$(\tau_2 \circ \Psi \circ \tau_1)_* = \Psi_*$$

at each point of the core submanifold of M_1. Hence by the uniqueness we have $\tau_2 \circ \Psi \circ \tau_1 = \Psi$. Therefore Ψ induces an isomorphism from $(M_1, \mathcal{F}_1)$ to $(M_2, \mathcal{F}_2)$. This completes the proof of Theorem 3.4.1. □

3.5. Isomorphisms and isometries

Let $(M, \mathcal{F})$ be a proper Liouville manifold of rank one, and let J_m $(1 \le m \le n-1)$ be as before. We first have the following

PROPOSITION 3.5.1. *The reflection with respect to J_m is a well-defined automorphism of the Liouville manifold $(M, \mathcal{F})$ for every m $(1 \le m \le n-1)$.*

We shall denote this reflection by τ_m.

PROOF. In the case of type (A) or (C) or (D), identify M with the quotient space $Q = R/G$ as described in Section 3.3 and define the mapping τ_m by

$$\tau_m(\Phi([x])) = \Phi([x_1, \dots, x_{m-1}, -x_m, x_{m+1}, \dots, x_n]).$$

Then as is easily seen, τ_m is a well-defined involution on Q and preserves the metric and $\mathcal{F}$. Also, it is clear that every point in J_m is fixed by τ_m and the normal vector there changes the sign by its differential. This implies that τ_m is the reflection with respect to the hypersurface J_m. In the case of type (B), we take the double covering, which is of type (A). Since τ_{m*} commutes with τ_*, the differential of the covering transformation at every point of the core submanifold, it follows that tau_m commutes with tau on the covering manifold. Hence we have the same result in this case. □

THEOREM 3.5.2. *Let $(M, \mathcal{F})$ be of type* (A) *with the core*

$$(\cap_m J_m, \{[f_1], \dots, [f_{n-1}]\}).$$

Assume that $[f_i] \neq [\cos^2(2\pi t/l) + c_i]$ for any $c_i \in \boldsymbol{R}$ and for every i, where $\cap_m J_m$ is identified with $\boldsymbol{R}/l\boldsymbol{Z}$ as in Proposition 3.1.3. Then $\mathcal{F}$ is unique. Namely, if $(M, \mathcal{F}')$ is another Liouville manifold of rank one defined over the same riemannian manifold M, then $\mathcal{F} = \mathcal{F}'$.

PROOF. For the 2 dimensional case, the uniqueness of $\mathcal{F}$ has been shown in [**Ki1**] Theorem 5.1 under the assumption that the curvature of M is not constant. An easy calculation shows that this condition is equivalent to $[f_1] \neq [\cos^2(2\pi t/l) + c_1]$ for any $c_1 \in \boldsymbol{R}$.

Now we assume $n \geq 3$. Let F_m, I_m, J_m be as before, and let F'_m, I'_m, J'_m be the corresponding objects for $(M, \mathcal{F}')$. We put $M^{0'} = M - \cup_m I'_m$. Since $\cap_{k\neq m} J_k$ is totally geodesic, it follows that each F'_i, restricted to $T^*(\cap_{k\neq m} J_k)$ still commutes with the energy function with respect to the Poisson bracket. Hence by the result for the 2 dimensional case, we have

$$F'_i = aF_m + 2bE \qquad \text{on} \quad T^*(\cap_{k\neq m} J_k), \tag{3.5.1}$$

where $a, b \in \boldsymbol{R}$. This implies that nonzero covector $\lambda \in T^*(\cap_k J_k)$ is an eigenvector of the endomorphisms F'^e_i at any point $p \in \cap_k J_k$. Hence it follows that if $p \notin I'_m$ for any m, then $T_p(\cap_k J_k)$ is admissible with respect to $(M, \mathcal{F}')$.

We claim that I'_m does not contain an open interval in $\cap_k J_k$ for any m. In fact, assume that it is not the case. Let p be a point on such an interval. Let w_1 and w_2 be an orthonormal basis of $N_p I'_m$ such that w_1 is normal to $T_p J'_m$. Let a_j and a be constants such that

$$G' = \sum_{j=1}^{n-1} a_j F'_j + 2aE$$

is equal to $w_1^2 + w_2^2$ at p. By the proof of Proposition 3.3.5 we can take G' so that G'_q is of rank 2 and positive semi-definite for any $q \in I'_m$.

Since w_1 is normal to $\cap_k J_k$, there is l such that the normal vector to J_l at p is not orthogonal to w_1. This implies that G' is not identically zero on $T^*_p(\cap_{k\neq l} J_k)$. Hence

$$G'|_{T^*(\cap_{k\neq l} J_k)}$$

is of rank one at p. Since its kernel is $T_p^*(\cap_k J_k)$, it follows that

$$G' = bF_l \qquad \text{on} \quad T^*(\cap_{k\neq l} J_k) \tag{3.5.2}$$

for some constant $b \neq 0$. Let L be the connected component of the intersection

$$J'_m \cap \cap_{k\neq l} J_k$$

through p. Since the intersection is transversal at p, and since both J'_m and $\cap_{k\neq l} J_k$ are closed and totally geodesic, it follows that L is a closed geodesic. Hence $L = \cap_k J_k$. Also, since $(F'_m)_q = 0$ for $q \in \cap_k J_k$ near p, we have $F'_m = 0$ on $T^*(\cap_{k\neq l} J_k)$ by (3.5.1). Since $T_q J'_m \not\supset T_q(\cap_{k\neq l} J_k)$ for any $q \in L$, this implies that $(F'_m)_q = 0$ for any $q \in L$. Hence we have $L \subset I'_m$. However, it contradicts the formula (3.5.2), because F_l takes both positive and negative values on $T^*(\cap_{k\neq l} J_k)|_L$. Thus we conclude that $M^{0'}$ contains an open and dense subset of $\cap_k J_k$.

Now let $p \in M^{0'} \cap \cap_k J_k$. Let w_k $(1 \leq k \leq n)$ be an admissible orthonormal basis of $T_p M$ with respect to $(M, \mathcal{F}')$ such that w_n is tangent to $T_p(\cap_k J_k)$. Then there are constants a_{ij} such that

$$G'_i = \sum_j a_{ij} F'_j = w_i^2 \qquad \text{at} \quad p.$$

We show that G'_i $(1 \leq i \leq n-1)$ is of rank one or zero at any point on $\cap_k J_k$. In fact, if G'_i is proportional to some F'_j at p, then $\cap_k J_k \subset J'_j$. Otherwise, let N' be the admissible submanifold determined by the orthogonal complement of $\boldsymbol{R}w_i$. Clearly it contains a neighborhood of p in $\cap_k J_k$. Assume that $N' \not\supset \cap_k J_k$, and let q be a boundary point of the connected component (closed interval) of $N' \cap_k J_k$ containing p. Then, at q the normal subspace to N' is admissible with respect to $(M, \mathcal{F}')$, and it is extended to an admissible subbundle P of TM on a neighborhood of q. This implies that $P_{q'}$ and $T_{q'}(\cap_k J_k)$ are mutually orthogonal for $q' \in M^{0'} \cap \cap_k J_k$ near q. Since $M^{0'}$ contains an open and dense subset of $\cap_k J_k$, we conclude that $P_{q'}$ and $T_{q'}(\cap_k J_k)$ are mutually orthogonal for every $q' \in \cap_k J_k$ near q. Since N' is locally an integral manifold of the orthogonal complement of P, it follows that N' contains a neighborhood of q in $\cap_k J_k$, a contradiction. Thus we have $N' \supset \cap_k J_k$, which implies that G'_i is of rank one at every point on $\cap_k J_k$.

Next we show that for each $p \in M^{0'} \cap \cap_k J_k$ an admissible basis with respect to $(M, \mathcal{F}')$ is also admissible with respect to $(M, \mathcal{F})$. Let G'_m and w_m $(1 \leq m \leq n)$ be as above. Fix i $(\leq n-1)$ and assume that there are m_1 and m_2 $(m_1 \neq m_2)$ such that w_i is not tangent to both J_{m_1} and J_{m_2}. Then we have

$$G'_i = c_j F_{m_j} \qquad \text{on} \quad T^*(\cap_{k\neq m_j} J_k), \tag{3.5.3}$$

where c_j are non-zero constants. Considering the signs of F_{m_j} on $\cap_k J_k$, we see that there is a point q on $\cap_k J_k$ such that the sign of $c_1 F_{m_1}$ and that of $c_2 F_{m_2}$ are mutually distinct. Since $(G'_i)_q$ is of rank one, we have a contradiction. Hence it follows that the basis $w_1, \ldots, w_n$ is also admissible with respect to $(M, \mathcal{F})$.

Also, in view of the formula (3.5.3) we see that each G'_i is of rank one or zero on $\cap_k J_k$ and has the same zero points as some F_j there. We claim that each G'_i $(1 \leq i \leq n-1)$ is equal to some F'_k. In fact, if it is not the case, then one can think of the maximal admissible submanifold N' with respect to $(M, \mathcal{F}')$ determined by the orthogonal complement of w_i. Then, as shown above, G'_i is of rank one at every point on $\cap_k J_k$. Since $(G'_i)_q = 0$ for some $q \in \cap_k J_k$, it is a contradiction. Hence G'_i is equal to some F'_k.

Consequently, we have seen that $\cap_k J_k = \cap_k J'_k$ and the identity mapping $\cap_k J_k \to \cap_k J'_k$ is an isomorphism of the cores of $(M, \mathcal{F})$ and $(M, \mathcal{F}')$. Also, the normal bundles NJ_i and NJ'_i, restricted to $\cap_k J_k$, coincide (by changing the subscripts of J'_i suitably). Therefore, by virtue of Proposition 3.4.2 we obtain an isomorphism

$$\Psi : (M, \mathcal{F}) \to (M, \mathcal{F}')$$

of Liouville manifolds such that Ψ_* is the identity on $TM|_{\cap_k J_k}$. Since $\Psi : M \to M$ is an isometry, the property above implies that Ψ is the identity on M. This completes the proof. □

The following corollary is clear.

COROLLARY 3.5.3. *Let $(M, \mathcal{F})$ and $(M', \mathcal{F}')$ be two proper Liouville manifolds of rank one and type* (A) *that satisfy the same assumption as in Theorem 3.5.2. Then any isometry $M \to M'$ is also an isomorphism of Liouville manifolds.*

We also have the following

COROLLARY 3.5.4. *Let $(M, \mathcal{F})$ be as in Theorem 3.5.2. Then the isometry group of M is finite.*

PROOF. Let ϕ be an isometry of M. Then by Theorem 3.5.2 we have $\phi_* \mathcal{F} = \mathcal{F}$. Hence ϕ becomes an automorphism of the Liouville manifold $(M, \mathcal{F})$. Let $\widehat{\phi}$ denote the induced automorphism of the core. By definition, an automorphism of the core is a diffeomorphism of the core submanifold that preserves the set of equivalence classes of functions on it. Identifying the core submanifold with $\boldsymbol{R}/l\boldsymbol{Z}$ as before, one can easily see that the automorphism group of the core is a subgroup of the group of diffeomorphisms (of order 8) generated by the translation by adding $l/4$ and the reflection with respect to 0.

Now, assume that $\widehat{\phi}$ is the identity. Then by Propositions 3.4.2 and 3.5.1, ϕ is identical with the composition of the reflections with respect to some J_i in this case. Hence the corollary follows. □

3.6. $C_{2\pi}$-metrics

In this section we shall describe a new family of $C_{2\pi}$-metrics on the n-sphere S^n. Let $a_1, \ldots, a_{n-1}$ be constants such that

$$0 < a_1 < \cdots < a_{n-1} < 1,$$

and put $\beta_i = \arcsin \sqrt{a_i} \in (0, \pi/2)$. Let $h(r)$ be a function on $\boldsymbol{R}/2\pi\boldsymbol{Z}$ that satisfies the following conditions:

(1) $h(-r) = -h(r)$ for $r \in [-\beta_1, \beta_1]$;
(2) $h(\pi - r) = -h(r)$ for $r \in [\beta_{n-1}, \pi - \beta_{n-1}]$;
(3) $h(\pi + r) = -h(r)$ for any $r \in \boldsymbol{R}/2\pi\boldsymbol{Z}$;
(4) $h(r) = 0$ for $r \in [\beta_1, \beta_{n-1}] \cup [\pi - \beta_{n-1}, \pi - \beta_1]$;
(5) $|h(r)| < 1$ for any $r \in \boldsymbol{R}/2\pi\boldsymbol{Z}$.

To the pair $(\{a_i\}, h(r))$ we assign a proper Liouville manifold of rank one and type (A) as follows. First we define the coordinate change $r \mapsto t = t(r)$ of $\boldsymbol{R}/2\pi\boldsymbol{Z}$ by the conditions $t(\beta_1) = \beta_1$ and

$$dt = (1 + h(r))dr.$$

The inverse mapping is denoted by $r(t)$. Let $f_i(t)$ $(1 \le i \le n-1)$ be functions on $\boldsymbol{R}/2\pi\boldsymbol{Z}$ given by

$$f_i(t) = a_i - \sin^2 r(t).$$

Clearly they satisfy the conditions (A.1), ..., (A.4) in Section 3.2 (with $f_i(\beta_i) = 0$). Let $(M, \mathcal{F})$ be the proper Liouville manifold of rank one and type (A) whose core is isomorphic to

$$(\boldsymbol{R}/2\pi\boldsymbol{Z}, \{[f_1], \dots, [f_{n-1}]\}).$$

We shall identify it with the constructed one and use the same notations as in Section 3.3. For instance points in M are expressed as

$$\Phi([x_1, \dots, x_n]), \qquad [x] \in \prod_{i=1}^{n} (\boldsymbol{R}/\alpha_i\boldsymbol{Z}).$$

PROPOSITION 3.6.1. *Every geodesic of M is closed and has the length 2π. In other words, M is a $C_{2\pi}$-manifold.*

REMARK. 1. M is isometric to the sphere of constant curvature 1 if and only if $h = 0$.

2. M is not isometric to any of Weinstein's $C_{2\pi}$-manifolds (cf. [**Be**] p. 120), because the isometry group of M is finite when $h \ne 0$.
3. Assume $h \ne 0$. Then two pairs $(\{a_i\}, h(r))$ and $(\{\widetilde{a}_i\}, \widetilde{h}(r))$ give mutually isometric Liouville manifolds if and only if $\widetilde{a}_i = a_i$ and $\widetilde{h}(r) = \pm h(\pm r)$, or $\widetilde{a}_i = 1 - a_{n-i}$ and $\widetilde{h}(r) = \pm h(\pi/2 \pm r)$.

They are immediate consequences of Theorem 3.5.2 and Corollary 3.5.4.

PROOF. Take $[x] \in \prod_i (\boldsymbol{R}/\alpha_i\boldsymbol{Z})$ and $\lambda \in S^*_{\Phi([x])}M$ so that $\Phi([x]) \in M^0$ and $\xi_i(\Phi^*\lambda) \ne 0$ for all i. Put $F_j(\lambda) = c_j$ $(1 \le j \le n-1)$. Remarking the formula

$$\sum_{j=1}^{n} b_{ij}\overline{F}_j = \xi_i^2, \tag{3.6.1}$$

where

$$\overline{F}_n = 2\overline{E}, \quad b_{ij} = \begin{cases} (-1)^i \prod_{k\ne j} \left(a_k - \sin^2 r(t_i(x_i))\right) & \text{if } j \le n-1 \\ (-1)^{i+1} \prod_{k=1}^{n-1} \left(a_k - \sin^2 r(t_i(x_i))\right) & \text{if } j = n, \end{cases}$$

we put
(3.6.2)

$$H(w) = \sum_{j=1}^{n-1} A_j(w)c_j + A_n(w), \quad A_j(w) = \begin{cases} \prod_{k\ne j}(a_k - w) & \text{if } 1 \le j \le n-1 \\ -\prod_{k=1}^{n-1}(a_k - w) & \text{if } j = n \end{cases}.$$

Then $H(w)$ is a polynomial of degree $n-1$ and satisfies

$$(-1)^i H\left(\sin^2 r(t_i(x_i))\right) > 0 \qquad (1 \le i \le n).$$

Hence there are constants $b_1, \dots, b_{n-1}$ such that

$$H(w) = -\prod_{j=1}^{n-1}(b_j - w) \tag{3.6.3}$$

$$\sin^2 r(t_i(x_i)) < b_i < \sin^2 r(t_{i+1}(x_{i+1})) \qquad (1 \le i \le n-1) \tag{3.6.4}$$

Note that $a_{i-1} \le \sin^2 r(t_i(x_i)) \le a_i$ for any $i \ge 2$. Hence

$$\begin{aligned} &a_{i-1} < b_i < a_{i+1} \qquad (2 \le i \le n-2), \\ &0 < b_1 < a_2, \quad a_{n-2} < b_{n-1} < 1 \\ &b_1 < \cdots < b_{n-1} \end{aligned} \tag{3.6.5}$$

Also, substituting $w = a_j$ in the formulas (3.6.2) and (3.6.3), we have

$$c_j = \frac{\prod_{k=1}^{n-1}(b_k - a_j)}{\prod_{k \ne j}(a_k - a_j)}. \tag{3.6.6}$$

Conversely, let $b_1, \dots, b_{n-1}$ be constants satisfying (3.6.5), and define $\{c_j\}$ by (3.6.6). Then it is easily seen that for any $[x]$ satisfying (3.6.4) there is $\lambda \in S^*_{\Phi([x])}M$ such that $F_i(\lambda) = c_i$ $(1 \le i \le n-1)$.

Now we fix $b_1, \dots, b_{n-1}$ satisfying (3.6.5) and consider the geodesics (integral curves of X_E) such that $F_i = c_i$ $(1 \le i \le n-1)$ and $2E = 1$. They satisfy the equations

$$\frac{dx_i}{ds} = \frac{\partial \overline{E}}{\partial \xi_i} \qquad (1 \le i \le n). \tag{3.6.7}$$

From the formula (3.6.1) we easily get

$$(b_{ij})^{-1} = \left(\frac{1}{2\xi_j}\frac{\partial \overline{F}_i}{\partial \xi_j}\right).$$

Hence using the formula (3.6.1) we have

$$\sum_{i=1}^{n} \frac{b_{ij}(x_i)}{\sqrt{(-1)^i H\left(\sin^2 r(t_i(x_i))\right)}} \left|\frac{dx_i}{ds}\right| = \begin{cases} 0 & \text{if } 1 \le j \le n-1 \\ 1 & \text{if } j = n. \end{cases} \tag{3.6.8}$$

Computing $(b_{ij})^{-1}$ explicitly, we also have

$$\frac{dx_i}{ds} = \frac{(-1)^{i-1}\xi_i}{\prod_{\substack{1 \le j \le n \\ j \ne i}} \left(\sin^2 r(t_j(x_j)) - \sin^2 r(t_i(x_i))\right)}. \tag{3.6.9}$$

From the formula (3.6.9) one can easily see the behavior of the geodesics: Each $x_i(s)$ oscillates in an interval, or moves on the whole circle with everywhere non-zero velocity, and the image coincides with a connected component of

$$U_i = \{x_i \in \boldsymbol{R}/\alpha_i \boldsymbol{Z} \mid \max\{a_{i-1}, b_{i-1}\} \le \sin^2 r(t_i(x_i)) \le \min\{a_i, b_i\}\}, \tag{3.6.10}$$

where we have put $a_0 = b_0 = 0$ and $a_n = b_n = 1$. Let m_i be the number of connected components of the set U_i. Then $m_i = 1$ if $b_{i-1} \le a_{i-1}$ and $a_i \le b_i$, $m_i = 2$ if $a_{i-1} < b_{i-1}$ and $a_i \le b_i$, or $b_{i-1} \le a_{i-1}$ and $b_i < a_i$, and $m_i = 4$ if $a_{i-1} < b_{i-1}$ and $b_i < a_i$.

In case $h = 0$, i.e., M is of constant curvature 1, every geodesic is closed and has the length 2π. In this case one can easily see that $x_i(s)$ oscillates in the interval described above with the least period $4\pi/m_i$ if $m_i \geq 2$, and that $x_i(s)$ moves monotonously on the whole circle $\boldsymbol{R}/\alpha_i\boldsymbol{Z}$ with the least period 2π if $m_i = 1$. Therefore integrating both sides of the formula (3.6.8) on the interval $0 \leq s \leq 2\pi$, we have

$$\text{(3.6.11)} \qquad \sum_{i=1}^{n} \int_{U_i} \frac{b_{ij}(x_i)}{\sqrt{(-1)^i H\left(\sin^2 r(t_i(x_i))\right)}}\, dx_i = \begin{cases} 0 & \text{if } 1 \leq j \leq n-1 \\ 2\pi & \text{if } j = n\,, \end{cases}$$

provided $h = 0$.

LEMMA 3.6.2. *The formula* (3.6.11) *is also valid for non-zero* h.

PROOF. Changing the variables in the integrals (3.6.11) we see that the left-hand side is equal to

$$\sum_i \int_{V_i} \frac{(-1)^i A_j(\sin^2 r)}{\sqrt{(-1)^i H(\sin^2 r)}} \frac{1 + h(r)}{\sqrt{(-1)^i A_n(\sin^2 r)}}\, dr\,,$$

where

$$V_i = \{r \in \boldsymbol{R}/2\pi\boldsymbol{Z} \mid \max\{a_{i-1}, b_{i-1}\} \leq \sin^2 r \leq \min\{a_i, b_i\}\}.$$

Since V_i are invariant under the involutions $r \mapsto -r$ and $r \mapsto \pi - r$, the integrals are equal to those in the case where $h = 0$. $\square$

Let $x^0 = (x_1^0, \ldots, x_n^0) \in \boldsymbol{R}^n$ satisfying (3.6.4) and

$$a_{i-1} < \sin^2 r(t_i(x_i^0)) < a_i.$$

LEMMA 3.6.3. *Put*

$$u_j = \sum_{i=1}^{n} \int_{x_i^0}^{y_i} \frac{b_{ij}(x_i)}{\sqrt{(-1)^i H\left(\sin^2 r(t_i(x_i))\right)}}\, dx_i.$$

Then the mapping $y = (y_1, \ldots, y_n) \mapsto u = (u_1, \ldots, u_n)$ *gives a diffeomorphism from a neighborhood of* x^0 *to a neighborhood of* $0 \in \boldsymbol{R}^n$.

The proof is easy, because $\det(b_{ij}(x_i^0)) \neq 0$.

Now we continue the proof of Proposition 3.6.1. First suppose that the C^0 norm of h is sufficiently small. Then by Lemmas 3.6.2 and 3.6.3 we have $\Phi([x(2\pi)]) = \Phi([x(0)])$. Next suppose that h is arbitrary. Replacing h with th $(0 \leq t \leq 1)$, we also have $\Phi([x(2\pi)]) = \Phi([x(0)])$ for small t. Since the set of such t is open and closed in $[0,1]$, we obtain the same result for h. The result implies that every geodesic with $F_i = c_i$ and $2E = 1$ is closed with period 2π. By moving $\{b_i\}$ so that the conditions (3.6.5) are satisfied, we see that $\zeta_{2\pi}\lambda = \lambda$ for λ in a dense subset of S^*M. Hence it follows that every geodesic is closed with period 2π. That 2π is in fact the least period follows from the fact that the geodesic flow of such M is symplectically isomorphic to that of the sphere of constant curvature 1 (cf. [**Be**] p. 122). $\square$

APPENDIX. SIMPLY CONNECTED MANIFOLDS OF CONSTANT CURVATURE

In this appendix we shall explain how the simply connected manifolds of constant curvature become Liouville manifolds of rank one in an extrinsic way. Also we shall concretely describe the branched coverings given in Section 3.3, which are necessary in the proof of Theorem 3.3.1.

A.1. Possible cores

Let $(M, \mathcal{F})$ be a proper Liouville manifold of rank one such that M is a simply connected riemannian manifold with the constant curvature K ($K = \pm 1, 0$). In this section we determine the core of such manifold. Let g_0 be the riemannian metric of M and let $(L, \{[f_1], \dots, [f_{n-1}]\})$ be the core of $(M, \mathcal{F})$. Then the core submanifold L is isometric to $\boldsymbol{R}/2\pi\boldsymbol{Z}$ if $K = 1$, and to $\boldsymbol{R}$ otherwise. Choosing the origin and the orientation of L suitably, we have the following

PROPOSITION A.1.1. *The representatives f_i can be taken of the form*

$$f_i(t) = f(t) + a_i \qquad (1 \le i \le n-1)$$

where t is the length parameter, a_i are constants, and:
(1) $f(t) = \cos 2t, \quad -1 < a_1 < \cdots < a_{n-1} < 1 \qquad (K = 1)$;
(2) $f(t) = -t^2, \quad 0 < a_1 < \cdots < a_{n-1} \qquad (K = 0$, type (C));
(3) $f(t) = -t, \quad a_1 < \cdots < a_{n-1} \qquad (K = 0$, type (D));
(4) $f(t) = -\cosh 2t, \quad 1 < a_1 < \cdots < a_{n-1} \qquad (K = -1$, type (C));
(5) $f(t) = -e^{2t}, \quad 0 < a_1 < \cdots < a_{n-1} \qquad (K = -1$, type (D));
(6) $f(t) = -\sinh 2t, \quad a_1 < \cdots < a_{n-1} \qquad (K = -1$, type (D)).

PROOF. We may assume that f_i satisfy the conditions in Propositions 3.2.1 and 3.2.2. Since the core of the two-dimensional Liouville manifold $\cap_{k \ne i} J_k$ is given by $(L, \{[f_i]\})$, it is enough to consider the two-dimensional case for the determination of each f_i. So let us assume that M is two-dimensional and let $(L, \{[f]\})$ be its core. Let $\beta = \beta_1$ be as in Propositions 3.2.1 and 3.2.2. Then in view of Section 3.3 the riemannian metric g_0 is expressed as

$$g_0 = (h_1(x_1) - h_2(x_2))(dx_1^2 + dx_2^2), \tag{A.1.1}$$

where $h_i(x_i) = f(t_i(x_i))$ and

$$\left(\frac{dt_i}{dx_i}\right)^2 = (-1)^{i-1} f(t_i),$$

$$t_1(0) = \beta, \quad \frac{dt_1}{dx_1}(0) < 0, \quad \text{etc..}$$

Computing the sectional curvature K we have

$$-(h_1'' - h_2'')(h_1 - h_2) + {h_1'}^2 + {h_2'}^2 = 2K(h_1 - h_2)^3. \tag{A.1.2}$$

Put $h_1''(0) = -h_2''(\alpha_2/2) = a$. The formula (A.1.2) implies

$$-(h_i'' + (-1)^{i-1}a)h_i + {h_i'}^2 = (-1)^{i-1} 2K h_i^3 \quad (i = 1, 2). \tag{A.1.3}$$

By (A.1.2) and (A.1.3) we have

$$\frac{h_2'' + a}{h_2} + \frac{h_1'' - a}{h_1} = -6K(h_1 - h_2). \tag{A.1.4}$$

Since K is constant, it then follows that

$$h_i'^2 = (-1)^{i-1}(-4Kh_i^3 + \frac{b}{a}h_i^2 + 2ah_i), \tag{A.1.5}$$

where $a = h_1''(0)$ and $b = h_1''''(0)$. The equations (A.1.5) turn into the single differential equation for $f(t)$:

$$f'^2 = -4Kf^2 + 2f''(\beta)f + f'(\beta)^2. \tag{A.1.4}$$

This equation is easily solved, and under the additional conditions for $f(t)$ described in Propositions 3.2.1 and 3.2.2 we see that $f(t)$ has one of the forms among (1), ..., (6) up to constant factor and a possible change of the length parameter t, i.e., $\pm t$ for the cases of type (A) and (C), and $\pm t + c$ ($c \in \boldsymbol{R}$) for the case of type (D).

To complete the proof of the proposition we have to verify that $f_i - f_j$ are constant in the case where $K = -1$ and the type is (D). For this it is enough to consider the 3-dimensional Liouville manifold $\cap_{k \neq i,j} J_k$. Then computing the sectional curvature of the 2-planes normal to the core submanifold one can easily see that $f_i - f_j$ is constant. □

A.2. The sphere S^n

Let E_{ij} be the matrix of size $(n+1) \times (n+1)$ such that the (i,j)-entry is 1 and others are 0. Put $X_{ij} = E_{ij} - E_{ji}$, which are elements of the Lie algebra $so(n+1)$. Let $b_1, \dots, b_{n+1}$ be arbitrary real constants such that

$$b_1 > \cdots > b_{n+1},$$

and fix them. Put

$$F(\lambda) = \sum_{i<j} \frac{X_{ij}^2}{(b_i - \lambda)(b_j - \lambda)} \in S^2(so(n+1)) \quad \text{(symmetric tensor product)},$$

where λ is a real parameter. Then it is directly verified that $\{F(\lambda), F(\mu)\} = 0$ for any $\lambda, \mu \in \boldsymbol{R}$, where $\{\ ,\ \}$ stands for the standard Poisson bracket on $C^\infty(so(n+1)^*)$. Hence describing

$$F(\lambda) = \sum_{i=1}^{n+1} \frac{F_{i-1}}{b_i - \lambda}, \quad F_{i-1} = \sum_{j \neq i} \frac{X_{ij}^2}{b_j - b_i},$$

we have $\{F_i, F_j\} = 0$ for any i, j. Note that $\sum_{i=1}^{n+1} F_{i-1} = 0$.

Let S^n be the unit sphere of

$$\boldsymbol{R}^n = \{(z_1, \dots, z_{n+1})\}$$

equipped with the induced metric g_0. The natural action of the group $SO(n+1)$ on S^n induces the homomorphism $so(n+1) \to \Gamma(TS^n)$ of the Lie algebras, and also the homomorphisms between their symmetric tensor products. We denote the images of $F(\lambda)$ and F_j with this homomorphism by the same symbols. We also denote by $\mathcal{F}$ the vector space spanned by $F_0, \dots, F_n$. Let E_0 denote the energy function associated with the metric g_0. Then

$$2E_0 = -\sum_{i=1}^{n+1} b_i F_{i-1} \in \mathcal{F}.$$

Also we have

$$F(\lambda) = \left(\sum_{i=1}^{n+1} \frac{z_i^2}{b_i - \lambda}\right)\left(\sum_{i=1}^{n+1} \frac{(\partial/\partial z_i)^2}{b_i - \lambda}\right) - \left(\sum_{i=1}^{n+1} \frac{z_i \partial/\partial z_i}{b_i - \lambda}\right)^2.$$

Let $\lambda_1, \dots, \lambda_n$ be the functions on

$$\overset{\circ}{S^n} = \{z \in S^n \mid z_i \neq 0 \text{ for any } i\}$$

given by the identity

$$\text{(A.2.1)} \qquad \sum_{i=1}^{n+1} \frac{z_i^2}{b_i - \lambda} = \frac{\prod_{k=1}^{n}(\lambda_k - \lambda)}{\prod_{i=1}^{n+1}(b_i - \lambda)}$$

and the inequalities $\lambda_1 > \cdots > \lambda_n$. Clearly $b_i > \lambda_i > b_{i+1}$ on $\overset{\circ}{S^n}$. Put

$$W_k = \sum_{i=1}^{n+1} \frac{z_i \partial/\partial z_i}{b_i - \lambda_k}.$$

As easily observed, $W_1, \dots, W_n$ form an orthogonal frame on $\overset{\circ}{S^n}$, and

$$d\lambda_i(W_j) = 2\delta_{ij}.$$

Since $F(\lambda_k) = -W_k^2$, it thus follows that $(S^n, \mathcal{F})$ is a Liouville manifold.

Now, let $2 \leq i \leq n$. On the great sphere defined by $z_i = 0$, F_{i-1} is written as

$$F_{i-1} = \left(\sum_{j \neq i} \frac{z_j^2}{b_j - b_i}\right)\left(\frac{\partial}{\partial z_i}\right)^2.$$

We put

$$J_{i-1} = \{z \in S^n \mid z_i = 0\},$$
$$I_{i-1} = \{z \in J_{i-1} \mid \sum_{j \neq i} \frac{z_j^2}{b_j - b_i} = 0\}.$$

Then we see that $\dim \mathcal{F}_z = n$ if and only if $z \notin \cup_{k=1}^{n-1} I_k$. Moreover, putting

$$X = \sum_{j \neq i} \frac{z_j(\partial/\partial z_j)}{b_j - b_i},$$

we see that X is normal to I_{i-1}, tangent to J_{i-1}, and

$$X \sum_{j \neq i} \frac{z_j^2}{b_j - b_i} \neq 0$$

at $z \in I_{i-1}$. These facts imply the properness of the Liouville manifold $(S^n, \mathcal{F})$ and that $F_{i-1} \in \mathcal{F}^*$ $(2 \leq i \leq n)$. Hence $(S^n, \mathcal{F})$ is a proper Liouville manifold of rank one and type (A). Parametrizing the core submanifold $\cap J_k$ by

$$z_1 = \sin t, \quad z_{n+1} = \cos t,$$

and describing $F_{i-1} = f_{i-1}(\partial/\partial z_i)^2$ there, one gets

$$f_{i-1} = \frac{-(b_1 - b_{n+1})}{2(b_1 - b_i)(b_i - b_{n+1})}\left(\cos 2t + \frac{b_1 + b_{n+1} - 2b_i}{b_1 - b_{n+1}}\right).$$

Putting $a_{i-1} = (b_1 + b_{n+1} - 2b_i)/(b_1 - b_{n+1})$, we see that the core is isomorphic to the one given in Proposition A.1.1 (1).

From the formula (A.2.1) we have

$$z_i^2 = \frac{\prod_{k=1}^n(\lambda_k - b_i)}{\prod_{j\neq i}(b_j - b_i)}. \tag{A.2.2}$$

Putting $\lambda_j = b_{j+1}\cos^2\theta_j + b_j \sin^2\theta_j$, and taking the square roots appropriately in (A.2.2), we obtain:

$$\begin{aligned} z_1 &= \cos\theta_1\sqrt{\prod_{j=2}^{n}\frac{b_1-\lambda_j}{b_1-b_{j+1}}} \\ z_i &= \sin\theta_{i-1}\cos\theta_i\sqrt{\prod_{j=1}^{i-2}\frac{\lambda_j-b_i}{b_j-b_i}\prod_{j=i+1}^{n}\frac{b_i-\lambda_j}{b_i-b_{j+1}}} \qquad (2\le i\le n) \\ z_{n+1} &= \sin\theta_n\sqrt{\prod_{j=1}^{n-1}\frac{\lambda_j-b_{n+1}}{b_j-b_{n+1}}}. \end{aligned} \tag{A.2.3}$$

The formulas above give the mapping

$$(\boldsymbol{R}/2\pi\boldsymbol{Z})^n \to S^n \qquad ((\theta_1,\dots,\theta_n)\mapsto z).$$

As easily seen, the inverse image of each point is a $\overline{G}$-orbit, where the group $\overline{G}$ of transformations of $(\boldsymbol{R}/2\pi\boldsymbol{Z})^n$ is generated by $\overline{\sigma}_1,\dots,\overline{\sigma}_{n-1}$;

$$\overline{\sigma}_j(\theta) = (\theta_1,\dots,\theta_{j-1},-\theta_j,\pi-\theta_{j+1},\theta_{j+2},\dots,\theta_n).$$

Now let us introduce new variables $x_1,\dots,x_n$ by the formulas:

$$x_j = x_j(\theta_j) = \int_0^{\theta_j}\frac{d\theta_j}{\sqrt{(-1)^{n-j}\prod_{k\neq j,j+1}(b_k - \lambda_j(\theta_j))}}.$$

Put $x_j(2\pi) = \alpha_j$. Then the change of variables $\theta\mapsto x$ gives the diffeomorphism

$$(\boldsymbol{R}/2\pi\boldsymbol{Z})^n \to R = \prod_{j=1}^{n}(\boldsymbol{R}/\alpha_j\boldsymbol{Z}),$$

and we have

$$g_0 = \sum_{j=1}^{n}(-1)^{n-j}\prod_{k\neq j}(\lambda_k - \lambda_j)\,dx_j^2. \tag{A.2.4}$$

By (A.2.2) and (A.2.3), one can easily see that the resulting mapping

$$R \to S^n \qquad (x\mapsto z)$$

induces a smooth bijective mapping $R/G \to S^n$, where the group G and the manifold structure of R/G are those as given in Section 3.3. Moreover, since the right-hand side of the formula (A.2.4) is non-degenerate on R/G, it follows that this induced mapping is actually a diffeomorphism. Also, it can be directly verified that R and the Liouville manifold $S^n = R/G$ are the same one as those constructed from the representatives $(b_1 - b_i)(b_{n+1} - b_i)f_{i-1}$ $(2 \le i \le n)$.

A.3. The euclidean space $\boldsymbol{R}^n$

Let $\boldsymbol{R}^n = \{(z_1, \dots, z_n)\}$ be the n-dimensional euclidean space with the standard flat metric g_0, and let $\mathfrak{g}$ the Lie algebra of the group of euclidean transformations of it. $\mathfrak{g}$ is naturally written as the semi-direct sum:

$$\mathfrak{g} = so(n) + \boldsymbol{R}^n,$$

whose translation part is generated by $T_j = \frac{\partial}{\partial z_j}$. First we shall describe the case of type (C). Let $b_1 > \cdots > b_n$ be real constants, and put

$$\begin{aligned} F(\lambda) &= \sum_{i>j} \frac{X_{ij}^2}{(b_i - \lambda)(b_j - \lambda)} - \sum_{k=1}^{n} \frac{T_k^2}{b_k - \lambda} \\ &= \sum_{i=1}^{n} \frac{F_{i-1}}{b_i - \lambda} \in S^2\mathfrak{g}. \end{aligned}$$

Then we have

$$\{F(\lambda), F(\mu)\} = 0 \qquad \text{for any} \quad \lambda, \mu \in \boldsymbol{R}.$$

As an element of $\Gamma(S^2T\boldsymbol{R}^n)$, $\sum_k T_k^2 = 2E_0 = -\sum_i F_i$, and

$$F(\lambda) = \left(\sum_i \frac{z_i^2}{b_i - \lambda} - 1\right)\left(\sum_i \frac{(\partial/\partial z_i)^2}{b_i - \lambda}\right) - \left(\sum_i \frac{z_i \partial/\partial z_i}{b_i - \lambda}\right)^2.$$

Let $\mathcal{F}$ be the vector space spanned by all F_i. Then one can easily see that $(\boldsymbol{R}^n, \mathcal{F})$ is a proper Liouville manifold of rank one, where $F_i \in \mathcal{F}^*$ $(1 \le i \le n-1)$. The core submanifold is given by

$$\{z \in \boldsymbol{R}^n \mid z_2 = \cdots = z_n = 0\},$$

on which F_i is written as

$$F_i = \left(\frac{z_1^2}{b_1 - b_{i+1}} - 1\right)\left(\frac{\partial}{\partial z_{i+1}}\right)^2.$$

Hence this Liouville manifold is of type (C), and putting $a_i = b_1 - b_{i+1}$, we see that its core is isomorphic to the one given in Proposition A.1.1 (2).

Now define the functions $\lambda_1, \dots, \lambda_n$ on

$$\overset{\circ}{\boldsymbol{R}^n} = \{z \in \boldsymbol{R}^n \mid z_i \ne 0 \ \text{ for every } i\}$$

by the formula

$$\sum_i \frac{z_i^2}{b_i - \lambda} - 1 = -\frac{\prod_k (\lambda_k - \lambda)}{\prod_i (b_i - \lambda)}$$

and the conditions $b_i > \lambda_i > b_{i+1}$, where $b_{n+1} = -\infty$. From the formula above we get

$$z_i^2 = -\frac{\prod_k(\lambda_k - b_i)}{\prod_{j\neq i}(b_j - b_i)}.$$

Putting $\lambda_j = b_{j+1}\cos^2\theta_j + b_j \sin^2\theta_j$ for $j \leq n-1$, $\lambda_n = b_n - \theta_n^2$, and taking the square root of z_i^2 appropriately, we obtain the mapping

$$(\boldsymbol{R}/2\pi\boldsymbol{Z})^{n-1} \times \boldsymbol{R} \to \boldsymbol{R}^n \qquad ((\theta_1, \dots, \theta_n) \mapsto z).$$

Next define the variables $x_1, \dots, x_n$ by

$$x_i = \int_0^{\theta_i} \frac{d\theta_i}{\sqrt{(-1)^{n-1-i}\prod_{j\neq i,i+1}(b_j - \lambda_i)}} \qquad (i \leq n-1),$$

$$x_n = \int_0^{\theta_n} \frac{d\theta_n}{\sqrt{\prod_{j\neq n}(b_j - \lambda_n)}},$$

and put $x_i(2\pi) = \alpha_i$ $(i \leq n-1)$, $x_n(\infty) = \alpha_n$. The coordinate change $\theta \mapsto x$ being a diffeomorphism, we have the mapping

$$R = \prod_{i=1}^{n-1}(\boldsymbol{R}/\alpha_i\boldsymbol{Z}) \times (-\alpha_n, \alpha_n) \to \boldsymbol{R}^n,$$

and the formula

$$g_0 = \sum_{i=1}^{n}(-1)^{n-i}\prod_{k\neq i}(\lambda_k - \lambda_i)\, dx_i^2.$$

Hence, as in the previous section, we have the diffeomorphism from the manifold R/G described in Section 3.3 to $\boldsymbol{R}^n$.

Next we shall describe the case of type (D). Let $\mathfrak{g}$, and $X_{ij}, T_i \in \mathfrak{g}$ be as before, and let $b_1 < \cdots < b_{n-1}$ be real constants. In this case we put

$$\begin{aligned} F(\lambda) &= \sum_{1\leq j<i\leq n-1} \frac{X_{ij}^2}{(b_i - \lambda)(b_j - \lambda)} - 2\sum_{i=1}^{n-1}\frac{T_i X_{ni}}{b_i - \lambda} + \lambda\sum_{i=1}^{n-1}\frac{T_i^2}{b_i - \lambda} - T_n^2 \\ &= \sum_{i-1}^{n-1}\frac{F_i}{b_i - \lambda} - \sum_{i=1}^{n} T_i^2. \end{aligned}$$

Then one gets

$$\{F(\lambda), F(\mu)\} = 0 \qquad \text{for any} \quad \lambda, \mu \in \boldsymbol{R}.$$

As a section of $S^2T\boldsymbol{R}^n$, $\sum_i T_i^2 = 2E_0$, and $F(\lambda)$ is expressed as

$$\left(\sum_{i=1}^{n-1}\frac{z_i^2}{b_i - \lambda} - 2z_n + \lambda\right)\left(\sum_{i=1}^{n-1}\frac{(\partial/\partial z_i)^2}{b_i - \lambda}\right) - \left(\sum_{i=1}^{n-1}\frac{z_i\partial/\partial z_i}{b_i - \lambda} - \partial/\partial z_n\right)^2.$$

Let $\mathcal{F}$ be the vector space spanned by E_0 and all F_i. Then as is easily seen, $(\boldsymbol{R}^n, \mathcal{F})$ becomes a proper Liouville manifold of rank one, where $F_i \in \mathcal{F}^*$. The core submanifold is given by

$$\{z \in \boldsymbol{R} \mid z_1 = \cdots = z_{n-1} = 0\},$$

on which F_i is written as

$$F_i = (-2z_n + b_i)\left(\frac{\partial}{\partial z_i}\right)^2.$$

Hence this Liouville manifold is of type (D), and putting $a_i = b_i/2$, we see that its core is isomorphic to the one given in Proposition A.1.1 (3).

Define the functions $\lambda_1, \ldots, \lambda_n$ by the formula

$$\sum_{i=1}^{n-1} \frac{z_i^2}{b_i - \lambda} - 2z_n + \lambda = -\frac{\prod_{i=1}^{n}(\lambda_i - \lambda)}{\prod_{i=1}^{n-1}(b_i - \lambda)}$$

and the conditions $b_{i-1} < \lambda_i < b_i$, where $b_0 = -\infty$ and $b_n = \infty$. We have

$$z_i^2 = -\frac{\prod_{k=1}^{n}(\lambda_k - b_i)}{\prod_{j\neq i}(b_j - b_i)} \quad (i \leq n-1), \quad z_n = \frac{1}{2}\left(\sum_{k=1}^{n}\lambda_k - \sum_{j=1}^{n-1} b_j\right)$$

Putting

$$\lambda_i = b_i\cos^2\theta_i + b_{i-1}\sin^2\theta_i \quad (2 \leq i \leq n-1), \quad \lambda_1 = b_1 - \theta_1^2, \ \lambda_n = b_{n-1} + \theta_n^2,$$

and taking the square roots of z_i^2 appropriately, we obtain the mapping

$$\boldsymbol{R} \times (\boldsymbol{R}/2\pi\boldsymbol{Z})^{n-2} \times \boldsymbol{R} \to \boldsymbol{R}^n \qquad ((\theta_1, \ldots, \theta_n) \mapsto z).$$

Now define the variables $x_1, \ldots, x_n$ by the formula:

$$x_k = \int_0^{\theta_k} \frac{d\theta_k}{\sqrt{(-1)^{k-2}\prod_{j\neq k-1,k}(b_j - \lambda_k)}} \qquad (2 \leq k \leq n-1)$$

$$x_1 = \int_0^{\theta_1} \frac{d\theta_1}{\sqrt{\prod_{j\geq 2}(b_j - \lambda_1)}}, \qquad x_n = \int_0^{\theta_n} \frac{d\theta_n}{\sqrt{(-1)^{n-2}\prod_{j\leq n-2}(b_j - \lambda_n)}},$$

and put $x_1(\infty) = \alpha_1$, $x_k(2\pi) = \alpha_k$ $(2 \leq k \leq n-1)$, $x_n(\infty) = \alpha_n$. Then one can easily see that the mapping

$$R = (-\alpha_1, \alpha_1) \times \prod_{k=2}^{n-1}(\boldsymbol{R}/\alpha_k\boldsymbol{Z}) \times (-\alpha_n, \alpha_n) \to \boldsymbol{R}^n \quad (x \mapsto z)$$

induces the diffeomorphism from the manifold R/G described in Section 3.3 to $\boldsymbol{R}^n$.

A.4. The hyperbolic space H^n

Let $\boldsymbol{R}^{1,n} = \{(z_0, z_1, \ldots, z_n)\}$ be the $(n+1)$-dimensional Minkowski space with the flat Lorentz metric $-dz_0^2 + dz_1^2 + \cdots + dz_n^2$, and let H^n be the hyperbolic space realized in $\boldsymbol{R}^{1,n}$ by the equation

$$-z_0^2 + \sum_{j=1}^{n} z_j^2 = -1, \qquad z_0 > 0.$$

Let $so(1,n)$ be the Lie algebra of the group of Lorentz transformations, which is spanned by

$$Y_j = E_{0j} + E_{j0} \qquad (1 \leq j \leq n)$$

and $X_{ij} = E_{ij} - E_{ji}$ $(1 \leq i, j \leq n)$. In this section we shall again start with the case of type (C). Let $b_0 > b_1 > \cdots > b_n$ be real constants, and put

$$F(\lambda) = -\sum_{i=1}^{n} \frac{Y_i^2}{(b_0 - \lambda)(b_i - \lambda)} + \sum_{i>j} \frac{X_{ij}^2}{(b_i - \lambda)(b_j - \lambda)}$$

$$= \sum_{i=0}^{n} \frac{F_{i-1}}{b_i - \lambda} \in S^2(so(1,n)). \tag{A.4.1}$$

Then we have

$$\{F(\lambda), F(\mu)\} = 0 \qquad \text{for any} \quad \lambda, \mu \in \boldsymbol{R}.$$

As a section of $S^2TH^n \subset S^2T\boldsymbol{R}^{1,n}$, $F(\lambda)$ is described as

$$F(\lambda) = \left(-\frac{z_0^2}{b_0 - \lambda} + \sum_{i=1}^{n} \frac{z_i^2}{b_i - \lambda}\right)\left(-\frac{(\partial/\partial z_0)^2}{b_0 - \lambda} + \sum_{i=1}^{n} \frac{(\partial/\partial z_i)^2}{b_i - \lambda}\right) - \left(\sum_{i=0}^{n} \frac{z_i \partial/\partial z_i}{b_i - \lambda}\right)^2.$$

We also have

$$\sum_{i=0}^{n} F_{i-1} = 0, \qquad \sum_{i=0}^{n} b_i F_{i-1} = 2E_0.$$

Let $\mathcal{F}$ be the vector space spanned by all F_i. Then we see that $(H^n, \mathcal{F})$ is a proper Liouville manifold of rank one, where $F_i \in \mathcal{F}^*$ $(1 \leq i \leq n-1)$. The core submanifold is given by

$$\{z \in H^n \mid z_2 = \cdots = z_n = 0\}.$$

parametrizing the core submanifold by

$$z_0 = \cosh t, \quad z_1 = \sinh t,$$

and describing $F_i = f_i(\partial/\partial z_i)^2$ $(1 \leq i \leq n-1)$ there, one gets

$$f_i = \frac{b_1 - b_0}{2(b_0 - b_{i+1})(b_n - b_{i+1})}\left(-\cosh 2t + \frac{b_0 + b_1 - 2b_{i+1}}{b_0 - b_1}\right).$$

Putting $a_i = (b_0 + b_1 - 2b_{i+1})/(b_0 - b_1)$, we see that the core is isomorphic to the one given in Proposition A.1.1 (4).

REMARK. If the constants $b_0, \ldots, b_n$ are taken so that

$$b_1 > \cdots > b_{i-1} > b_0 > b_i > \cdots > b_n \qquad (2 \leq i \leq n)$$

in the formula (A.4.1), then one can see that $F_j \in \mathcal{F}^*$ if and only if $0 \leq j < i-2$ or $i \leq j \leq n-1$. Consequently the Liouville manifold $(H^n, \mathcal{F})$ in this case becomes of rank two.

Next we shall describe the case of type (D). Let $b_1 > \cdots > b_{n-1}$ be real constants, and put

$$F(\lambda) = -Y_n^2 + \lambda \sum_{i=1}^{n-1} \frac{Y_i^2 - X_{ni}^2}{b_i - \lambda} + \sum_{i=1}^{n-1} \frac{(Y_i - X_{ni})^2}{b_i - \lambda} + \frac{\lambda^2}{2} \sum_{i,j=1}^{n-1} \frac{X_{ij}^2}{(b_i - \lambda)(b_j - \lambda)}$$

$$= -\sum_{i=1}^{n} Y_i^2 + \frac{1}{2} \sum_{i,j=1}^{n} X_{ij}^2 + \sum_{i=1}^{n-1} \frac{F_i}{b_i - \lambda}$$

$$G(\lambda) = -Y_n^2 + \lambda \sum_{i=1}^{n-1} \frac{Y_i^2 - X_{ni}^2}{b_i - \lambda} - 2 \sum_{i=1}^{n-1} \frac{Y_i X_{ni}}{b_i - \lambda} + \frac{\lambda^2 + 1}{2} \sum_{i,j=1}^{n-1} \frac{X_{ij}^2}{(b_i - \lambda)(b_j - \lambda)}$$

$$= -\sum_{i=1}^{n} Y_i^2 + \frac{1}{2} \sum_{i,j=1}^{n} X_{ij}^2 + \sum_{i=1}^{n-1} \frac{G_i}{b_i - \lambda}$$

In the case of $F(\lambda)$ the assumption " $b_1 < 0$ " is also necessary.

Then one gets

$$\{F(\lambda), F(\mu)\} = \{G(\lambda), G(\mu)\} = 0 \qquad \text{for any} \quad \lambda, \mu \in \boldsymbol{R}.$$

As a section of S^2TH^n, we have the formula

$$-\sum_{i=1}^{n} Y_i^2 + \frac{1}{2} \sum_{i,j=1}^{n} X_{ij}^2 = -2E_0.$$

Also $F(\lambda)$, $G(\lambda)$ are expressed as

$$\begin{aligned}\lambda^2 F(\lambda) = &\left((z_0 + z_n)^2 + \lambda(z_0^2 - z_n^2) + \lambda^2 \sum_{i=1}^{n-1} \frac{z_i^2}{b_i - \lambda} \right) \\ &\times \left((\partial/\partial z_0 - \partial/\partial z_n)^2 + \lambda((\partial/\partial z_0)^2 - (\partial/\partial z_n)^2) + \lambda^2 \sum_{i=1}^{n-1} \frac{(\partial/\partial z_i)^2}{b_i - \lambda} \right) \\ &- \left(-(z_0 + z_n)(\partial/\partial z_0 - \partial/\partial z_n) - \lambda(z_0\partial/\partial z_0 + z_n\partial/\partial z_n) + \lambda^2 \sum_{i=1}^{n-1} \frac{z_i\partial/\partial z_i}{b_i - \lambda} \right)^2,\end{aligned}$$

$$\begin{aligned}(\lambda^2 + 1)G(\lambda) = &\left(2z_0 z_n + \lambda(z_0^2 - z_n^2) + (\lambda^2 + 1) \sum_{i=1}^{n-1} \frac{z_i^2}{b_i - \lambda} \right) \\ &\times \left(-2(\partial/\partial z_0)(\partial/\partial z_n) + \lambda((\partial/\partial z_0)^2 - (\partial/\partial z_n)^2) + (\lambda^2 + 1) \sum_{i=1}^{n-1} \frac{(\partial/\partial z_i)^2}{b_i - \lambda} \right) \\ &- \left(z_0\partial/\partial z_n - z_n\partial/\partial z_0 - \lambda(z_0\partial/\partial z_0 + z_n\partial/\partial z_n) + (\lambda^2 + 1) \sum_{i=1}^{n-1} \frac{z_i\partial/\partial z_i}{b_i - \lambda} \right)^2.\end{aligned}$$

Let $\mathcal{F}$ (resp. $\mathcal{G}$) be the vector space spanned by all F_i (resp. all G_i) and E_0. Then both $(H^n, \mathcal{F})$ and $(H^n, \mathcal{G})$ become proper Liouville manifolds of rank one, where $F_i \in \mathcal{F}^*$, $G_i \in \mathcal{G}^*$. The core submanifolds are given by

$$\{z \in \boldsymbol{R} \mid z_1 = \cdots = z_{n-1} = 0\}$$

in both cases. parametrizing the core submanifolds by

$$z_0 = \cosh t, \quad z_n = \sinh t,$$

and describing $F_i = f_i(\partial/\partial z_i)^2$, $G_i = g_i(\partial/\partial z_i)^2$, $(1 \leq i \leq n-1)$ there, one gets

$$f_i = e^{2t} + b_i, \qquad g_i = \sinh 2t + b_i$$

Putting $a_i = -b_i$ in both cases, we see that those cores are isomorphic to the one given in Proposition A.1.1 (5) and (6) respectively.

Part 2. Kähler-Liouville Manifolds

Introduction

It is known that the geodesic flow of the complex projective space CP^n equipped with the standard Kähler metric is (completely) integrable in the sense of symplectic geometry (or in Liouville's sense) (cf. Thimm [**Th**], see also [**Mi**], [**MF1**], [**IW**]). The first integrals are given as follows: Let $c_0, \dots, c_n$ be constants such that $1 = c_0 > c_1 > \cdots > c_n = 0$, and let $(z_0, \dots, z_n)$ be the homogeneous coordinate system. Then, by putting $\partial_i = \partial/\partial z_i$,

$$\sum_{\substack{0 \le j \le n \\ j \ne i}} \frac{(z_i \bar{\partial}_j - z_j \bar{\partial}_i)(\bar{z}_i \partial_j - \bar{z}_j \partial_i)}{c_j - c_i} \quad (1 \le i \le n-1),$$

$$\sum_{i,j} (z_i \bar{\partial}_j - z_j \bar{\partial}_i)(\bar{z}_i \partial_j - \bar{z}_j \partial_i), \quad \sqrt{-1}(z_i \partial_i - \bar{z}_i \bar{\partial}_i) \quad (1 \le i \le n)$$

are well-defined symmetric tensor fields and vector fields on CP^n. Regarded as functions on the cotangent bundle T^*CP^n, they become a system of first integrals in involution of the geodesic flow. (Note that the first integrals given here are slightly different from those in [**Th**].)

In this part, we shall define a class of Kähler manifolds whose geodesic flows are integrable by a set of first integrals possessing similar properties to those for CP^n, and study the properties of such manifolds. We shall call them Kähler-Liouville manifolds (of type (A)). The precise definition is as follows. Let M be an n-dimensional complete Kähler manifold, I its complex structure, and E its energy function (the hamiltonian of the geodesic flow). Let $\mathcal{F}$ be an n-dimensional vector space of sections of S^2TM (the symmetric tensor product over $\boldsymbol{R}$ of two copies of the tangent bundle TM). Then we say that $(M, \mathcal{F})$ is Kähler-Liouville manifold if it satisfies the following conditions:

(1) $E \in \mathcal{F}$;
(2) $\{F, H\} = 0$ for any $F, H \in \mathcal{F}$;
(3) every $F \in \mathcal{F}$ is "hermitian";
(4) F_p ($F \in \mathcal{F}$) are simultaneously normalizable for every $p \in M$;
(5) $\mathcal{F}_p = \{F_p \mid F \in \mathcal{F}\}$ is n-dimensional at some $p \in M$.

Here, $F_p \in S^2T_pM$ is the value of F at $p \in M$. Since sections of S^2TM are naturally regarded as functions on the cotangent bundle T^*M, the Poisson bracket in (2) makes sense. Also, (3) means that the function F, restricted to each fibre T_p^*M, is a hermitian form. We say that two Kähler-Liouville manifolds $(M, \mathcal{F})$ and $(M', \mathcal{F}')$ are mutually isomorphic if there is an isomorphism $\phi : M \to M'$ of Kähler manifolds that maps $\mathcal{F}$ to $\mathcal{F}'$.

As is immediately seen, only n first integrals are given in the definition. Nevertheless, it will turn out that other n first integrals appear automatically if some non-degeneracy condition is assumed. A Kähler-Liouville manifold satisfying this assumption is called *of type* (A) (for the precise definition, see Section 1). The main purpose of this part is to investigate local and global properties of Kähler-Liouville manifolds of type (A). Compact, 2-dimensional Kähler-Liouville manifolds were studied by Igarashi [I], in which he adopted another type of non-degeneracy condition and obtained similar results to ours in this case. The results indicate that the condition adopted in [I] is equivalent to our condition of type (A).

We now explain the various results in this part. In the following, Kähler-Liouville manifolds are assumed to be of type (A), unless otherwise stated.

1 (PROPOSITION 1.9). *A finite, partially ordered set $\mathcal{A}$ is naturally associated with $(M, \mathcal{F})$. Also, a positive integer $|\alpha|$ is assigned to each $\alpha \in \mathcal{A}$ so that $\sum_{\alpha \in \mathcal{A}} |\alpha| = n$. For any $\alpha \in \mathcal{A}$, the subset $\{\beta \in \mathcal{A} \mid \beta \prec \alpha\}$ is totally ordered.*

2 (PROPOSITION 1.16, THEOREMS 3.1, 3.2). *An n-dimensional commutative Lie algebra $\mathfrak{k}$ of infinitesimal automorphisms of $(M, \mathcal{F})$ is naturally defined, possessing the property that $\mathfrak{k}$ and $\mathcal{F}$ are elementwise commutative with respect to the Poisson bracket. With $\mathfrak{k}$ and $\mathcal{F}$ the geodesic flow of M becomes integrable.*

Up to now, further results are obtained only for compact Kähler-Liouville manifolds. In this case, we obtain the results below. Put $\mathfrak{g} = \mathfrak{k} + I\mathfrak{k}$, and let K and G be the transformation group of M generated by $\mathfrak{k}$ and $\mathfrak{g}$ respectively.

3 (THEOREMS 4.2, 4.18). *K and C are isomorphic to $U(1)^n$ and $(C^{\times})^n$ respectively. With the action of G, M becomes a toric variety.*

The structure of M as toric variety is completely investigated in Section 4, and as a consequence we obtain the notion of "toric variety of KL-A type" (cf. Section 5). The toric variety associated with a Kähler-Liouville manifold of type (A) is necessarily of KL-A type. Conversely, we have the following result.

4 (THEOREM 8.4). *Suppose that a toric variety of KL-A type is given. Then there exists a Kähler-Liouville manifold of type* (A) *such that the associated toric variety is isomorphic to the given one.*

Thanks to the general theory for toric varieties (cf. [**Da**], [**De**], [**Fu**], [**O1**]), we can know what kind of complex manifold M is. For the detail, see Section 5. In particular, we have the following bundle structures.

5 (PROPOSITION 5.4). *There is a holomorphic principal fibre bundle*

$$\prod_{\alpha \in \mathcal{A}} (C^{|\alpha|+1} - \{0\}) \to M$$

whose structure group is isomorphic to $(C^{\times})^{\#\mathcal{A}}$.

6 (PROPOSITION 5.5, THEOREMS 6.3, 6.4, 6.5). *Let $\mathcal{A}'$ be a subset of $\mathcal{A}$ possessing the property that if $\alpha \in \mathcal{A}'$ and $\beta \prec \alpha$, then $\beta \in \mathcal{A}'$. Put $\mathcal{A}'' = \mathcal{A} - \mathcal{A}'$. Then, associated with $\mathcal{A}'$, there naturally exist Kähler-Liouville manifolds $(M', \mathcal{F}')$, $(M'', \mathcal{F}'')$, and a holomorphic fibre bundle $\pi : M \to M''$ whose typical fibre is M'. They possess the following properties:* (1) *$(M'', \mathcal{F}'')$ is of type* (A) *and the associated partially ordered set is $\mathcal{A}''$;* (2) *if $(M', \mathcal{F}')$ is of type* (A), *then the associated partially*

ordered set is $\mathcal{A}'$; (3) there is a homomorphism $G \to G''$ so that $\pi : M \to M''$ is equivariant, where G and G'' denote the algebraic tori acting on M and M'' respectively; (4) even if $(M', \mathcal{F}')$ is not of type (A), *M' possess the structure of toric variety inherited from M so that the maximal compact subgroup of the algebraic torus acting on M' preserves the metric and each element of $\mathcal{F}'$.*

The property (4) stated above indicates that the geodesic flow of $(M', \mathcal{F}')$ is integrable even if it is not of type (A). In this case we shall say that the Kähler-Liouville manifold $(M', \mathcal{F}')$ is of type (B) (cf. Section 6). Such a manifold will be necessary for the study of Kähler-Liouville manifold of type (A) only when $\dim M' = 1$. Using the result above successively, we consequently obtain a family of Kähler-Liouville manifolds $(M_\alpha, \mathcal{F}_\alpha)$ $(\alpha \in \mathcal{A})$ such that the partially ordered set associated with $(M_\alpha, \mathcal{F}_\alpha)$ consists of one element $\{\alpha\}$. In this case, it turns out that M_α is isomorphic to $CP^{|\alpha|}$ as toric variety. It also turns out that it is of type (A) if and only if $|\alpha| \geq 2$.

The result 4 mentioned above is actually given in much finer form in Theorem 8.4. There, besides the toric variety M, we prescribe the Kähler-Liouville manifolds $(M_\alpha, \mathcal{F}_\alpha)$ $(\alpha \in \mathcal{A})$ and some constants, and prove the uniqueness of $(M, \mathcal{F})$ as well as the existence. The reason for formulating the "existence theorem" in this form is that Kähler-Liouville manifolds such that the associated partially ordered sets consist of one element are easily understandable by using the results in Part 1 of this paper. The result is roughly stated as follows.

7 (THEOREM 7.2). *The isomorphism classes of Kähler-Liouville manifolds such that $\#\mathcal{A} = 1$ are completely classified by means of several constants and a function on a circle.*

We now briefly explain the organization of Part 2. In Section 1 we first formulate some non-degeneracy condition (depending on points) on a Kähler-Liouville manifold $(M, \mathcal{F})$. Denoting by M^1 the set of points where the condition is satisfied, we say that $(M, \mathcal{F})$ is of type (A) if $M^1 \neq \emptyset$. We perform local calculus on M^1, and introduce almost all basic quantities related to $(M, \mathcal{F})$. In Sections 2 and 3 we sum up the local data given in Section 1, and describe properties of the basic quantities and the Lie algebra $\mathfrak{k}$ in global form. The result 2 stated above is proved in Section 3.

Through Sections 4–8 we assume that M is compact. In Section 4 we prove the result 3 stated above and the results that determine the structure of M as toric variety. In particular, we show that M^1 is the unique open G-orbit, and $M - M^1$ is the union of $n + \#\mathcal{A}$ closed hypersurfaces that are G-invariant and totally geodesic.

In Section 5 we describe the various properties of the toric variety M. There we specify the fan of M, and give the definition of toric variety of KL-A type. This section contains 3 subsections; "The fan of M", "Fibre bundles associated with M", and "Line bundles". In Section 6 we prove the result 5 stated above. There we also prove its converse (Theorem 6.11), which plays a crucial role in the proof of Theorem 8.4.

Section 7 is devoted to the proof of Theorem 7.2 (see the result 7 above). We establish the one-to-one correspondence between the isomorphism classes of Kähler-Liouville manifolds of type (A) such that $\#\mathcal{A} = 1$ and the isomorphism classes of Liouville manifolds of rank one and type (B) that satisfy a certain condition. The definition and the classification of Liouville manifolds of rank one are given in Part

1 of this paper. By using them, the theorem is proved. In Section 8 we prove Theorem 8.4 mentioned above. It is no longer hard by virtue of Theorems 6.11 and 7.2.

Preliminary remarks and notations

Let M be an n-dimensional Kähler manifold, and let g and I be its Kähler metric and complex structure respectively. Then the Kähler form ω and the energy function E are given as follows:

$$\omega(X,Y) = g(IX,Y) \quad (X,Y \in T_pM, p \in M) \qquad E(\lambda) = \frac{1}{2}|\lambda|^2 \quad (\lambda \in T^*M),$$

where $|\cdot|$ denotes the norm function on T^*M associated with the metric g. The energy function E is the hamiltonian of the geodesic flow of M. For a function h on M we define vector fields $\operatorname{grad} h$ and $\operatorname{sgrad} h$ by the following formulae:

$$i_{\operatorname{grad} h}g = dh, \qquad i_{\operatorname{sgrad} h}\omega = -dh,$$

where i denotes the inner derivation. We have $\operatorname{sgrad} h = I \operatorname{grad} h$.

Let $p \in M$, and let S^2T_pM be the symmetric tensor product of two copies of the tangent space T_pM. Let F be an element of S^2T_pM, and suppose that, regarded as a quadratic form on T_p^*M, F is a hermitian form. Then there is an orthonormal basis V_j, IV_j $(1 \le j \le n)$ of T_pM and constants $a_1, \dots, a_n$ such that

$$F = \sum_j a_j(V_j^2 + (IV_j)^2).$$

We define the endomorphism F^e of T_pM by

$$F^e(V_i) = a_iV_i, \quad F^e(IV_i) = a_iIV_i.$$

Clearly it is independent of the choice of $\{V_i\}$. Regarding T_pM as a complex vector space (by identifying I with $\sqrt{-1}$), F^e becomes $\boldsymbol{C}$-linear. We define $\operatorname{tr} F$ and $\det F$ by the trace and the determinant of F^e over $\boldsymbol{C}$ respectively ($\operatorname{tr} F = \sum_j a_j$, $\det F = \prod_j a_j$).

Let $(M, \mathcal{F})$ be a Kähler-Liouville manifold. Then the condition (4) in the definition is equivalent to that $\{F_p^e \mid F \in \mathcal{F}\}$ is commutative with respect to the composition of endomorphisms for every $p \in M$.

We shall use the term "smooth" in the same meaning as "of class C^∞".

1. Local calculus on M^1

Let $(M, \mathcal{F})$ be a Kähler-Liouville manifold of dimension n. Put

$$M^0 = \{p \in M \mid \dim \mathcal{F}_p = n\} \qquad M^s = M - M^0.$$

Then M^0 is an open subset of M, which is not empty because of the condition (5) in the definition of Kähler-Liouville manifold. Let $F_1, \dots, F_n$ be a basis of $\mathcal{F}$, and let p be a point of M^0. Then there are an orthonormal frame V_i, IV_i $(i = 1, \dots, n)$ and n^2 functions f_{ij} around p such that

$$F_i = \sum_{j=1}^n f_{ij}\ (V_j^2 + (IV_j)^2).$$

Putting $(a_{ij}) = (f_{ij})^{-1}$, we have

$$\sum_{j=1}^{n} a_{ij}F_j = V_i^2 + (IV_i)^2.$$

Let D_i be the subbundle of TM defined around p spanned by V_i and IV_i.

PROPOSITION 1.1. *There are positive functions $a_1, \dots, a_n$ around the point p such that, putting*

$$b_{ij} = \frac{a_{ij}}{a_i}, \qquad W_i = \frac{V_i}{\sqrt{a_i}},$$

$$W_j b_{ik} = (IW_j) b_{ik} = 0 \quad (j \neq i, \ \textit{any } k), \tag{1.1}$$

$$\{W_i^2 + (IW_i)^2, W_j^2 + (IW_j)^2\} = 0 \quad (\textit{any } i, j). \tag{1.2}$$

The function a_i can be chosen to be one of $|a_{i1}|, \dots, |a_{in}|$ that is non-zero around p. Moreover, if $\{a'_i\}$ possess the same properties as above, then

$$V_j \frac{a'_i}{a_i} = (IV_j)\frac{a'_i}{a_i} = 0 \quad (j \neq i).$$

PROOF. We have

$$\begin{aligned} \{V_i^2 + (IV_i)^2, V_j^2 + (IV_j)^2\} &= \sum_{k,l=1}^{n} (\{a_{ik}, F_l\} a_{jl} F_k + \{F_k, a_{jl}\} a_{ik} F_l) \\ &= \sum_k \{a_{ik}, V_j^2 + (IV_j)^2\} F_k + \sum_l \{V_i^2 + (IV_i)^2, a_{jl}\} F_l \end{aligned} \tag{1.3}$$

Note that each term in the formula above is a homogeneous polynomial of degree 3 in the variables V_k, IV_k $(1 \leq k \leq n)$. Since the left-hand side belongs to the ideal $(V_iV_j, V_iIV_j, V_jIV_i, IV_iIV_j)$ of the polynomial algebra, and since F_k are linear combinations of $V_l^2 + (IV_l)^2$ $(1 \leq l \leq n)$, it follows that

$$\sum_k \{a_{ik}, V_j\} F_k = c_{ij}(V_i^2 + (IV_i)^2), \quad \sum_k \{a_{ik}, IV_j\} F_k = d_{ij}(V_i^2 + (IV_i)^2) \tag{1.4}$$

for some functions c_{ij} and d_{ij}, provided $i \neq j$. Hence we have

$$\{a_{ik}, V_j\} = c_{ij} a_{ik}, \quad \{a_{ik}, IV_j\} = d_{ij} a_{ik}. \tag{1.5}$$

Let a_i be one of the functions $|a_{i1}|, \dots, |a_{in}|$ that does not vanish around the point p. Then by (1.5) we have

$$\left\{\frac{a_{ik}}{a_i}, V_j\right\} = \left\{\frac{a_{ik}}{a_i}, IV_j\right\} = 0 \qquad (i \neq j).$$

This implies that

$$\sum_k \{a_{ik}, V_j^2 + (IV_j)^2\} F_k = \frac{1}{a_i}\{a_i, V_j^2 + (IV_j)^2\}(V_i^2 + (IV_i)^2).$$

Hence, by (1.3) we obtain

$$\{\frac{1}{a_i}(V_i^2 + (IV_i)^2), \frac{1}{a_j}(V_j^2 + (IV_j)^2)\} = 0.$$

The remaining part is clear. □

PROPOSITION 1.2. $[V_i, IV_i] \equiv \operatorname{sgrad}(\log a_i) \mod D_i$.

PROOF. We use the Kähler form ω. We have

$$0 = d\omega(V_i, IV_i, V_j) = -\omega([V_i, IV_i], V_j) + \omega([V_i, V_j], IV_i) - \omega([IV_i, V_j], V_i).$$

Since (1.2) implies

$$\tag{1.6} \begin{aligned} &[W_i, W_j] = \alpha_{ij} IW_i - \alpha_{ji} IW_j, \quad [IW_i, IW_j] = \beta_{ij} W_i - \beta_{ji} W_j, \\ &[W_i, IW_j] = -\beta_{ij} IW_i + \alpha_{ji} W_j, \quad [IW_i, W_j] = -\alpha_{ij} W_i + \beta_{ji} IW_j \end{aligned}$$

for some functions α_{ij} and β_{ij} ($i \neq j$), it follows that

$$\omega([V_i, V_j], IV_i) = -\omega([IV_i, V_j], V_i) = -V_j \log \sqrt{a_i}.$$

Hence we have

$$\tag{1.7} \omega([V_i, IV_i], V_j) = -V_j \log a_i,$$

provided $j \neq i$. Replacing V_j with IV_j, we also have

$$\tag{1.8} \omega([V_i, IV_i], IV_j) = -IV_j \log a_i.$$

From (1.7) and (1.8) it thus follows that

$$[V_i, IV_i] \equiv \sum_{j \neq i} (-(IV_j \log a_i) V_j + (V_j \log a_i) IV_j) \mod D_i.$$

□

We now consider the following condition for points on M^0:

(1.9) For any i there is some j such that $da_j|_{D_i} \neq 0$ at p.

Note that this condition is independent of the choice of $\{a_i\}$. Put

$$M^1 = \{p \in M^0 \mid (1.9) \text{ holds at } p\}.$$

We shall say that a Kähler-Liouville manifold $(M, \mathcal{F})$ is *of type (A)* if

$$M^1 \neq \emptyset.$$

From now on (until the end of Section 4) Kähler-Liouville manifolds are assumed to be of type (A), unless otherwise stated.

Let $p \in M^1$.

PROPOSITION 1.3. *Let $j, k \neq i$. If $d\log a_j|_{D_i} \neq 0$ and $d\log a_k|_{D_i} \neq 0$ at $q \in M^1$ near p, then they are linearly dependent at q.*

PROOF. Note first that $\sum_j a_{jk}$ is constant for every k, because $E \in \mathcal{F}$. Since $a_{jk} = a_j b_{jk}$, this implies that $\{a_i\}$ and $\{a_{ij}\}$ are written as rational functions of $\{b_{ij}\}$. Hence for each i there is some l such that $db_{il} \neq 0$ around p. On the other hand, the kernel of $d\log a_j$ on D_i is spanned by

$$-(IV_i \log a_j)V_i + (V_i \log a_j)IV_i = [V_j, IV_j]_{D_i},$$

where the right-hand side denotes the D_i-component of $[V_j, IV_j]$. Since $V_j b_{il} = IV_j b_{il} = 0$, we also have

$$0 = [V_j, IV_j]b_{il} = [V_j, IV_j]_{D_i} b_{il}.$$

Hence the kernel of $d\log a_j$ on D_i coincides with that of db_{il} on D_i. Since the latter does not depend on j, the proposition follows. □

By virtue of the proposition above we can take the orthonormal frame V_i, IV_i $(i = 1, \dots, n)$ around $p \in M^1$ so that

$$d\log a_j(IV_i) = 0 \qquad \text{for any } j \neq i.$$

Note that V_i are uniquely determined up to sign (and the numbering). We shall assume that V_i are taken in this way. Let D^+ (resp. D^-) be the subbundle of TM spanned by $V_1, \dots, V_n$ (resp. $IV_1, \dots, IV_n$). D^+ and D^- are well-defined over M^1;

$$TM^1 = D^+ \oplus D^-.$$

PROPOSITION 1.4. (1) *da_i, da_{ij}, db_{ij} are zero on D^-.*
(2) *For any i there is some j such that $W_i b_{ij} \neq 0$.*
(3) *$[W_i, W_j] = [IW_i, IW_j] = [W_i, IW_j] = 0$ $(i \neq j)$. In particular, D^+ and D^- are integrable.*

PROOF. (1) and (2) are clear from the proof of Proposition 1.3. Suppose that $W_i b_{ik} \neq 0$. Then by (1.6),

$$0 = [IW_i, IW_j]b_{ik} = \beta_{ij} W_i b_{ik}, \quad 0 = [IW_i, W_j]b_{ik} = -\alpha_{ij} W_i b_{ik}.$$

Hence $\alpha_{ij} = \beta_{ij} = 0$, and (3) follows. □

PROPOSITION 1.5. $[W_i, IW_i] \in D^-$.

PROOF. By virtue of Propositions 1.2 and 1.4, it suffices to prove that the D_i-component of $[W_i, IW_i]$ belongs to D^-. Choose $j(\neq i)$ such that $d\log a_j|_{D_i} \neq 0$. Then the D_i-component of $[W_j, IW_j]$ is not zero. Describing

$$[W_j, IW_j] = \alpha W_j + \beta IW_j + \sum_{k \neq j} \gamma_k IW_k,$$

we have

$$0 = [W_i, [W_j, IW_j]]_{D_i} = (W_i \gamma_i) IW_i + \gamma_i [W_i, IW_i]_{D_i}.$$

Since $\gamma_i \neq 0$, the proposition follows. □

PROPOSITION 1.6. *Maximal integral manifolds of D^+ are (locally) totally geodesic.*

PROOF. Since $IW_k < W_i, W_j >= 0$ and $[W_i, IW_k] \in D^-$, it follows that

$$< \nabla_{W_i} W_j, IW_k >= 0$$

for any i, j, k, where ∇ denotes the Levi-Civita covariant derivative. Hence the proposition follows. □

By virtue of Proposition 1.5 the D_i-component of $[W_i, IW_i]$ is of the form $c_i IW_i$, c_i being the function around p.

PROPOSITION 1.7. (1) *For i, j such that $i \neq j$ and $W_i a_j \neq 0$,*

$$c_i = -W_i \log a_i + W_i \log a_j - \frac{W_i^2 \log a_j}{W_i \log a_j}.$$

(2) $(IW_k)c_i = 0 \qquad (any\ k)$.
(3) $W_k c_i = -(W_k \log a_i)(W_i \log a_k) \qquad (k \neq i)$.
(4) $W_j W_i \log a_i = (W_i \log a_j)(W_j \log a_i) \qquad (i \neq j)$.
(5) $W_i W_j \log a_k = 0 \qquad (i \neq j \neq k \neq i)$.
(6) $(W_i \log a_k)(W_j \log a_k) = (W_i \log a_j)(W_j \log a_k) + (W_j \log a_i)(W_i \log a_k)$ $(i \neq j \neq k \neq i)$.

PROOF. By Propositions 1.2 and 1.4 we have

$$[W_i, IW_i] = c_i IW_i + \sum_{j \neq i} \frac{a_j}{a_i}(W_j \log a_i) IW_j.$$

Then, computing $[W_k, [W_i, IW_i]]$ for $k \neq i$, we obtain

$$\begin{aligned} 0 &= (W_k c_i + (W_k \log a_i)(W_i \log a_k)) IW_i \\ &+ (c_k \frac{a_k}{a_i} W_k \log a_i + W_k(\frac{a_k}{a_i} W_k \log a_i)) IW_k \\ &+ \sum_{j \neq i,k} \left(W_k(\frac{a_j}{a_i} W_j \log a_i) + \frac{a_j}{a_i}(W_k \log a_i)(W_j \log a_k) \right) IW_j. \end{aligned}$$

This formula implies (1), (3), and

$$\begin{aligned} W_k W_j \log a_i &= (W_k \log a_i)(W_j \log a_i) \\ &- (W_k \log a_j)(W_j \log a_i) - (W_j \log a_k)(W_k \log a_i) \end{aligned} \tag{1.10}$$

for mutually distinct i, j, k. Also, (2) follows from (1) and Proposition 1.4. To prove (4), (5), and (6), we recall that $\sum_i a_i b_{il}$ are constants. Differentiating these by W_j and W_k successively $(j \neq k)$ we have

$$\sum_i (W_j \log a_i) a_i b_{il} + a_j W_j b_{jl} = 0,$$

$$\begin{aligned} &\sum_i (W_k W_j \log a_i) a_i b_{il} + (W_j \log a_k) a_k W_k b_{kl} \\ &+ \sum_i (W_j \log a_i)(W_k \log a_i) a_i b_{il} + (W_k \log a_j) a_j W_j b_{jl} = 0. \end{aligned}$$

Thus,

$$\begin{aligned} W_k W_j \log a_i &= (W_j \log a_k)(W_k \log a_i) \\ &+ (W_k \log a_j)(W_j \log a_i) - (W_k \log a_i)(W_j \log a_i). \end{aligned} \tag{1.11}$$

From (1.10) and (1.11) the formulae (5) and (6) follows. Also, since i is arbitrary in the formula (1.11), (4) follows by putting $i = k$ in (1.11). □

We now define the binary relation $\preceq$ on the set of indices $\{1, \dots, n\}$:

$$i \preceq j \Longleftrightarrow i \neq j \text{ and } W_i \log a_j \neq 0 \text{ at } p \in M^1, \text{ or } i = j.$$

When it is necessary to clarify the point-dependence, we shall say that $i \preceq j$ at p. Also, we write $i \sim j$ if $i \preceq j$ and $j \preceq i$.

LEMMA 1.8. (1) *If $i \preceq j$ and $j \preceq k$, then $i \preceq k$.*
(2) *The relation $\sim$ is an equivalence relation.*

PROOF. Assume that $i \not\preceq k$ and $j \neq i, k$. Then, by Proposition 1.7 (6) we have

$$(W_i \log a_j)(W_j \log a_k) = 0.$$

Hence $i \not\preceq j$ or $j \not\preceq k$, and (1) follows. (2) is an immediate consequence of (1). □

Let $\mathcal{A}$ be the set of the equivalence classes. It is clear from Lemma 1.8 (1) that the relation $\preceq$ induces a binary relation (denoted by the same symbol) on the set $\mathcal{A}$. For $\alpha \in \mathcal{A}$, let $|\alpha|$ denote the number of indices contained in the equivalence class α.

PROPOSITION 1.9. *Let $\alpha, \beta, \gamma \in \mathcal{A}$.*
(1) *If $\alpha \preceq \beta$ and $\beta \preceq \gamma$, then $\alpha \preceq \gamma$. Namely, the relation $\preceq$ is a partial order on $\mathcal{A}$.*
(2) *If $\alpha \preceq \gamma$ and $\beta \preceq \gamma$, then $\alpha \preceq \beta$ or $\beta \preceq \alpha$. Namely, for any fixed γ, the set of $\delta \in \mathcal{A}$ such that $\delta \preceq \gamma$ is a totally ordered subset of $\mathcal{A}$.*
(3) *If α is a maximal element, then $|\alpha| \geq 2$.*

PROOF. (1) is clear from Lemma 1.8 (1).
(2) Let $i \in \alpha$, $j \in \beta$, and $k \in \gamma$, and assume that $\alpha \not\preceq \beta$ and $\beta \not\preceq \alpha$. Then $W_i \log a_j = W_j \log a_i = 0$. Hence by Proposition 1.7 (6), we have either $W_i \log a_k = 0$ or $W_j \log a_k = 0$.
(3) Suppose α is maximal and $i \in \alpha$. Then by the condition (1.9) we see that there is some $j (\neq i)$ such that $W_i \log a_j \neq 0$. Since α is maximal, it follows that $j \in \alpha$. □

Let W_i^* $(1 \leq i \leq n)$ be 1-forms such that $W_i^*(IW_j) = 0$ and $W_i^*(W_j) = \delta_{ij}$. Then Propositions 1.4 and 1.5 imply that $dW_i^* = 0$. Hence there is a system of functions $(x_1, \dots, x_n)$ such that $dx_i = W_i^*$. Clearly, a_i are functions of $(x_1, \dots, x_n)$, and $W_j \log a_i$ is nothing but the derivative of $\log a_i$ with respect to the variable x_j. For simplicity, we shall write ∂_j instead of $\partial / \partial x_j$.

LEMMA 1.10. *For any i, j $(i \neq j)$, there are functions $h_{ij}(x_i)$ such that $h_{ij} - h_{ji} \neq 0$ at p, and*

$$\partial_i \log a_j = -\partial_i \log |h_{ij} - h_{ji}|.$$

Moreover, if $\partial_i \log a_j \neq 0$ and $\partial_i \log a_k \neq 0$ at p $(j, k \neq i)$, then there are constants $c (\neq 0)$ and d such that

$$h_{ik} = c h_{ij} + d.$$

PROOF. By Proposition 1.7 (5) the function $\partial_i \log a_j$ depends only on the variables x_i and x_j. Since

$$\partial_j \partial_i \log a_j = \partial_i \partial_j \log a_i = (\partial_i \log a_j)(\partial_j \log a_i),$$

it follows that there is a positive function $H = H(x_i, x_j)$ such that

$$\partial_i \log a_j = \partial_i \log H, \quad \partial_j \log a_i = \partial_j \log H, \quad \partial_i \partial_j \log H = (\partial_i \log H)(\partial_j \log H).$$

The last formula implies that

$$\partial_i \partial_j \left(\frac{1}{H}\right) = 0.$$

Hence there are functions $h_{ij}(x_i)$ and $h_{ji}(x_j)$ such that $H^{-1} = |h_{ij} - h_{ji}|$.

Now, let us assume that $\partial_i \log a_j \neq 0$ and $\partial_i \log a_k \neq 0$ at p. Then the derivatives h'_{ij} and h'_{ik} do not vanish at p. By virtue of Proposition 1.7 (1), the function

$$\partial_i \log a_j - \frac{\partial_i^2 \log a_j}{\partial_i \log a_j} = -\frac{h''_{ij}}{h'_{ij}}$$

does not depend on j. Hence it follows that

$$\frac{h''_{ij}}{h'_{ij}} = \frac{h''_{ik}}{h'_{ik}},$$

which proves the latter half of the lemma. □

PROPOSITION 1.11. *For any $p \in M^1$ there is a neighborhood U of p such that the relation $\preceq$ is stable on U. Namely,*

$$i \preceq j \ \text{at}\ p \Longleftrightarrow i \preceq j \ \text{at}\ q$$

for any $q \in U$.

PROOF. Let $U(\subset M^1)$ be a connected neighborhood of p such that every $\partial_i \log a_j$ $(i \neq j)$ that does not vanish at p also does not vanish everywhere on U, and that all the functions h_{ij} are defined there.

Now, let us assume that $i \not\preceq j$ at p and $i \preceq j$ at some $q \in U$. Let $k(\neq i)$ be a number such that $i \preceq k$ at p. Then $i \preceq k$ at q, and by Lemma 1.10,

$$h'_{ij} = c h'_{ik} \tag{1.12}$$

on a neighborhood of q for some constant $c \neq 0$. Let U' be the set of all $q' \in U$ such that the formula (1.12) is valid at q'. Clearly, U' is closed in U. Moreover, if $q' \in U'$, then $h'_{ij} \neq 0$ at q'. This implies $i \preceq j$ at q' by Lemma 1.10, and thus (1.12) holds on a neighborhood of q'. Hence U' is open. Since $q \in U'$ and U is connected, it follows that $U' = U$. But since $p \notin U'$, it is a contradiction. □

We shall write $\alpha \prec \beta$ $(\alpha, \beta \in \mathcal{A})$ if $\alpha \preceq \beta$ and $\alpha \neq \beta$.

PROPOSITION 1.12. *Let $p \in M^1$, and let U be a small neighborhood of p. Then there are functions h_i $(1 \le i \le n)$ on U and constants $e_{\alpha\beta}$ $(\alpha, \beta \in \mathcal{A}, \alpha \prec \beta)$ satisfying the following four conditions:*

(1) $dh_i|_{D^-} = dh_i|_{D_j} = 0 \quad (i \neq j), \qquad W_i h_i \neq 0$ *everywhere on* U;
(2) $e_{\alpha\beta} = e_{\alpha\gamma} \qquad (\alpha \prec \beta \preceq \gamma)$;
(3) $h_i \neq h_j \quad (i \sim j,\ i \neq j), \quad h_i + e_{\alpha\beta} \neq 0 \quad (i \in \alpha, \alpha \prec \beta)$ *everywhere on* U;
(4) *the functions a_i can be taken in the form*

$$a_i = \left| \prod_{\substack{j \in \alpha \\ j \neq i}} (h_j - h_i) \right|^{-1} \left| \prod_{\gamma \prec \alpha} \prod_{k \in \gamma} (h_k + e_{\gamma\alpha}) \right|^{-1} \qquad (i \in \alpha).$$

If $\{\widetilde{h}_i\}$ and $\{\widetilde{e}_{\alpha\beta}\}$ also satisfy the conditions above, then there are constants $c_\alpha \neq 0$ and d_α such that

$$\widetilde{h}_i = c_\alpha h_i - d_\alpha, \quad \widetilde{e}_{\alpha\beta} = c_\alpha e_{\alpha\beta} + d_\alpha \qquad (i \in \alpha).$$

PROOF. We determine the functions h_i and the constants $e_{\alpha\beta}$ inductively with respect to the partial order $\prec$. Let $h_{ij}(x_i)$ $(i \neq j)$ and U be as in Lemma 1.10 and Proposition 1.11 respectively.

Let $\alpha \in \mathcal{A}$ be a minimal element. Then the functions a_i $(i \in \alpha)$ depend only on the variables x_j $(j \in \alpha)$. Hence, if $|\alpha| = 1$, then we can put $a_i = 1$ (cf. Proposition 1.1). In this case the function h_i does not appear yet. Also, if $|\alpha| = 2$, then putting $h_i = h_{ij}$ and $h_j = h_{ji}$ $(i, j \in \alpha)$, we can define a_i and a_j as

$$a_i = a_j = \frac{1}{|h_i - h_j|}.$$

Now, suppose that $|\alpha| \ge 3$. By Lemma 1.10 we know that the ratios h'_{ij}/h'_{ik} are non-zero constants for any mutually distinct $i, j, k \in \alpha$. Hence there is a function $h_i(x_i)$ and non-zero constants c_{ij} such that

$$h'_{ij} = c_{ij} h'_i$$

for any $i, j \in \alpha$ $(i \neq j)$. Then by Proposition 1.7 (6) we have

$$c_{ij} c_{jk} c_{ki} = c_{ji} c_{kj} c_{ik} \qquad (i, j, k \in \alpha)$$

Namely, $\{c_{ij}/c_{ji}\}$ satisfies the cocycle condition. An easy calculation shows that it is actually a coboundary, i.e., there are non-zero constants c_i $(i \in \alpha)$ such that $c_{ij}/c_{ji} = c_i/c_j$. Then, putting

$$\widetilde{h}_{ij} = \frac{c_i}{c_{ij}} h_{ij}, \qquad \widetilde{h}_i = c_i h_i,$$

we obtain

$$\partial_i \widetilde{h}_{ij} = \partial_i \widetilde{h}_i, \qquad \widetilde{h}_{ij} - \widetilde{h}_{ji} = \frac{c_i}{c_{ij}} (h_{ij} - h_{ji}).$$

Putting $\widetilde{h}_{ij} - \widetilde{h}_i = d_{ij}$ (a constant), and using Proposition 1.7 (6) again, we have

$$d_{ij} + d_{jk} + d_{ki} = d_{ji} + d_{kj} + d_{ik}.$$

Put

$$d_i = |\alpha|^{-1} \sum_{\substack{j\in\alpha \\ j\neq i}} (d_{ij} - d_{ji}) \qquad (i \in \alpha).$$

Then we have

$$d_{ij} - d_{ji} = d_i - d_j, \quad \widetilde{h}_{ij} - \widetilde{h}_{ji} = (\widetilde{h}_i + d_i) - (\widetilde{h}_j + d_j)$$

for any $i, j \in \alpha$ $(i \neq j)$. Hence, by redefining h_i as

$$h_i = \widetilde{h}_i + d_i \qquad (i \in \alpha),$$

the function

$$\log a_i + \sum_{\substack{j\in\alpha \\ j\neq i}} \log |h_j - h_i| \qquad (i \in \alpha)$$

depends only on the single variable x_i. Thus we can take a_i $(i \in \alpha)$ as

$$a_i = \prod_{\substack{j\in\alpha \\ j\neq i}} |h_j - h_i|^{-1}.$$

Now, let α be a non-minimal element of $\mathcal{A}$, and let $\beta \in \mathcal{A}$ $(\beta \prec \alpha)$ be a unique element such that there is no $\gamma \in \mathcal{A}$ satisfying $\beta \prec \gamma \prec \alpha$. We assume that the functions h_i are defined for all $i \in \gamma$ and $\gamma \prec \alpha$ $(\gamma \prec \beta$ if $|\beta| = 1)$ and the constants $e_{\gamma\delta}$ are defined for all γ and δ $(\gamma \prec \delta \prec \alpha)$ so that the conditions (1), (2), (3), and (4) in the proposition are satisfied. Under this assumption we shall define suitable functions h_i $(i \in \alpha)$ when $|\alpha| \geq 2$, h_j $(j \in \beta)$ when $|\beta| = 1$, and a constant $e_{\beta\alpha}$.

Let $i \in \alpha$, $j \in \beta$, and $k \in \gamma(\preceq \beta)$. Then the assumption implies that

$$\partial_k \log a_j = \begin{cases} -\partial_k \log |h_k - h_j| & (\gamma = \beta,\ k \neq j) \\ -\partial_k \log |h_k + e_{\gamma\beta}| & (\gamma \prec \beta). \end{cases}$$

Since

$$(\partial_j \log a_i)(\partial_k \log a_i) = (\partial_j \log a_k)(\partial_k \log a_i) + (\partial_k \log a_j)(\partial_j \log a_i),$$

we have $\partial_k \log a_i = \partial_k \log a_j$ in case $\gamma \prec \beta$. Hence in this case, putting $e_{\gamma\alpha} = e_{\gamma\beta}$, we have

$$\partial_k \log |a_i(h_k + e_{\gamma\alpha})| = 0. \tag{1.13}$$

Also, in case $\gamma = \beta$ and $j \neq k$ (hence $|\beta| \geq 2$), we have

$$-h_k + \frac{h'_k}{h'_{ki}}(h_{ki} - h_{ik}) = -h_j + \frac{h'_j}{h'_{ji}}(h_{ji} - h_{ij}).$$

Since the left- and the right-hand side of the formula above are functions of x_k and x_j respectively, it follows that they are constants. Let $e_{\beta\alpha}$ be this common constant. Note that this constant does not depend on the choice of $i \in \alpha$. In fact, let $l \in \alpha$ $(l \neq i)$. Then by Proposition 1.7 (6) we have $\partial_j \log a_i = \partial_j \log a_l$, i.e.,

$$\frac{-h'_{ji}}{h_{ji} - h_{ij}} = \frac{-h'_{jl}}{h_{jl} - h_{lj}}. \tag{1.14}$$

Then we have

$$h_j + e_{\beta\alpha} = \frac{h'_j}{h'_{ji}}(h_{ji} - h_{ij}) \tag{1.15}$$

for any $j \in \beta$ and $i \in \alpha$, provided $|\beta| \geq 2$. Then, by (1.13) and (1.15) we see that

$$\partial_k \log \left(a_i \prod_{\gamma \prec \alpha} \prod_{l \in \gamma} |h_l + e_{\gamma\alpha}| \right) = 0 \tag{1.16}$$

for any $k \in \gamma$, $\gamma \prec \alpha$, and $i \in \alpha$, if $|\beta| \geq 2$. When $|\beta| = 1$, we put

$$h_j = |h_{ji} - h_{ij}|, \qquad e_{\beta\alpha} = 0$$

for $j \in \beta$ and a fixed $i \in \alpha$. Then by (1.14) we also have (1.16) in this case.

Now, applying the same argument as for minimal element, we obtain functions h_l $(l \in \alpha)$ so that the function

$$\log \left(a_i \prod_{\gamma \prec \alpha} \prod_{l \in \gamma} |h_l + e_{\gamma\alpha}| \right) + \log \prod_{\substack{l \in \alpha \\ l \neq i}} |h_l - h_i|$$

depends only on x_i for every $i \in \alpha$ (if $|\alpha| = 1$, the second term does not appear, and h_i $(i \in \alpha)$ remains undetermined). Hence a_i $(i \in \alpha)$ can be taken so that this function is equal to zero. This completes the induction. The remaining part of the proposition is easy. □

We now assume that each equivalence class $\alpha \in \mathcal{A}$ consists of successive numbers:

$$\alpha = \{s(\alpha), s(\alpha) + 1, \ldots, t(\alpha)\}.$$

Let $m(i)$ $(i \in \alpha)$ be the number of functions h_j $(j \in \alpha)$ such that $h_i > h_j$ at p. Also, let $m(\alpha, \beta)$ be the number of negative functions in

$$\{h_i + e_{\alpha\beta} \mid i \in \alpha\}.$$

Let $\mathfrak{n}(\alpha)$ be the set of $\beta \in \mathcal{A}$ such that $\alpha \prec \beta$ and there is no γ between α and β. Also, put

$$u_\alpha = \prod_{\gamma \prec \alpha} \prod_{l \in \gamma} (h_l + e_{\gamma\alpha}).$$

PROPOSITION 1.13. *For a suitably chosen basis $F_1, \ldots, F_n$ of $\mathcal{F}$, the functions $b_{ij} = a_{ij}/a_i$ $(i \in \alpha)$ are given by:*

$$b_{ij} = \begin{cases} (-1)^{m(i)}(-h_i)^{j-s(\alpha)} & (j \in \alpha) \\ (-1)^{m(i)-1+m(\alpha,\beta)}(h_i + e_{\alpha\beta})^{-1} & (j = t(\beta),\ \beta \in \mathfrak{n}(\alpha)) \\ 0 & (\textit{otherwise}). \end{cases}$$

Moreover,

$$\sum_{\beta \succeq \alpha} \sum_{j \in \beta} a_j b_{jk} = \begin{cases} |u_\alpha|^{-1} & (k = t(\alpha)) \\ 0 & (k \neq t(\alpha)). \end{cases}$$

PROOF. Put

$$\widetilde{b}_{ij} = \begin{cases} (-1)^{m(i)}(-h_i)^{j-s(\alpha)} & (j \in \alpha) \\ (-1)^{m(i)-1+m(\alpha,\beta)}(h_i + e_{\alpha\beta})^{-1} & (j = t(\beta),\ \beta \in \mathfrak{n}(\alpha)) \\ 0 & (\text{otherwise}). \end{cases}$$

Then a direct computation shows that

$$\sum_{\beta \succeq \alpha} \sum_{j \in \beta} a_j \widetilde{b}_{jk} = \begin{cases} |u_\alpha|^{-1} & (k = t(\alpha)) \\ 0 & (k \neq t(\alpha)). \end{cases}$$

Let B be the $n \times n$ matrix $(\widetilde{b}_{ij})$, and let B_i be the $(n-1) \times (n-1)$ matrix obtained by deleting i-th row and $t(\alpha)$-th column from B, where $i \in \alpha$.

LEMMA 1.14. (1) $\det B \neq 0$.
(2) $\det B_i \neq 0$.

PROOF. It is easily seen that $\det B$ is equal to $\prod_{\alpha \in \mathcal{A}} \det B^\alpha$, where $B^\alpha = (\widetilde{b}_{ij})_{i,j \in \alpha}$. Since $\det B^\alpha \neq 0$ by Vandermonde's formula, (1) follows. (2) is similar.□

Define functions c_{kj} $(1 \leq k, j \leq n)$ by the formula

$$b_{ij} = \sum_k \widetilde{b}_{ik} c_{kj}. \tag{1.17}$$

To prove Proposition 1.13 it suffices to show that c_{kj} are constants. Let $\alpha \in \mathcal{A}$ and $l \in \alpha$. We claim that $\partial_l c_{t(\alpha),j} = 0$ for any j. In fact, since $\sum_i a_i b_{ij}$ are constants, we obtain

$$0 = \partial_l \sum_i a_i b_{ij} = \partial_l \sum_{\beta \succeq \alpha} \sum_{i \in \beta} a_i b_{ij}.$$

By (1.17) we also have

$$\sum_{\beta \succeq \alpha} \sum_{i \in \beta} a_i b_{ij} = |u_\alpha|^{-1} c_{t(\alpha),j}.$$

Thus, it follows that $\partial_l c_{t(\alpha),j} = 0$.

Moreover, the formula (1.16) implies that

$$0 = \sum_{k \neq t(\alpha)} \widetilde{b}_{ik} \partial_l c_{kj},$$

provided $i \neq l$ and $l \in \alpha$. Hence by Lemma 1.14 we have

$$\partial_l c_{kj} = 0 \qquad (l \in \alpha,\ k \neq t(\alpha)).$$

This completes the proof. □

Correspondingly, $(f_{ij}) = (a_{ij})^{-1}$ is given as follows.

PROPOSITION 1.15. *Suppose $i \in \alpha$. Then:*

$$f_{ij} = \begin{cases} |u_\alpha| \mathcal{S}_{t(\alpha)-i}(h_l\,;\, l \in \alpha - \{j\}) & (j \in \alpha) \\ |u_\alpha| \sum_{m=0}^{t(\alpha)-i} e_{\alpha\beta}^m \mathcal{S}_{t(\alpha)-i-m}(h_l\,;\, l \in \alpha) & (j \in \gamma \succeq \beta, \beta \in \mathfrak{n}(\alpha)) \\ 0 & (\textit{otherwise}), \end{cases}$$

where $\mathcal{S}_m(h_l\,;\, l \in \alpha)$ stands for the elementary symmetric function of degree m $(0 \leq m \leq |\alpha|)$ with respect to $|\alpha|$ functions $\{h_l \mid l \in \alpha\}$.

The proof is straightforward. Put

$$v_i = u_\alpha \mathcal{S}_{t(\alpha)-i+1}(h_l\,;\, l \in \alpha) \qquad (i \in \alpha,\ \alpha \in \mathcal{A}),$$

and let $\mathcal{V}$ be the vector space of functions on U spanned by constant functions and v_i $(1 \leq i \leq n)$. Put

$$\mathfrak{k} = \{\mathrm{sgrad}\,(v) \mid v \in \mathcal{V}\}.$$

PROPOSITION 1.16.
(1) $[Y, W_j] = [Y, IW_j] = 0$ $\quad(Y \in \mathfrak{k},$ *any* $j)$.
(2) $\{Y, F\} = 0$ $\quad(Y \in \mathfrak{k},\ F \in \mathcal{F})$.
(3) *$\mathfrak{k}$ is a commutative Lie algebra of infinitesimal automorphisms of the Kähler manifold M on U.*

PROOF. Put $Y = \mathrm{sgrad}\, v_m$. Then

$$Y = \sum_j a_j (W_j v_m) I W_j.$$

Hence, by Propositions 1.4 and 1.5 we have $[Y, IW_j] = 0$ for any j. Since

$$[W_i, IW_i] = \left(-\partial_i \log a_i - \frac{h_i''}{h_i'}\right) IW_i + \sum_{j \neq i} \frac{a_j}{a_i} (\partial_j \log a_i) IW_j,$$

it follows that

$$[W_i, Y] = a_i \left(\partial_i^2 v_m - (\partial_i v_m) \frac{h_i''}{h_i'}\right) IW_i$$
$$+ \sum_{j \neq i} a_j (\partial_i \partial_j v_m + (\partial_i \log a_j)(\partial_j v_m) + (\partial_j \log a_i)(\partial_i v_m)) IW_j.$$

Then, it is easily seen that each term in the right-hand side of the formula above vanishes. Thus (1) follows.

From (1) it follows that

$$0 = \{Y, W_i^2 + (IW_i)^2\} = \sum_j b_{ij} \{Y, F_j\}.$$

This indicates (2). In particular, we have $\{Y, E\} = 0$, which implies that Y is an infinitesimal isometry. The property (1) also implies that $(\mathcal{L}_Y I) W_i = (\mathcal{L}_Y I) IW_i = 0$ for any i, where $\mathcal{L}_Y$ denotes the Lie derivative with respect to Y. Hence it follows that $\mathcal{L}_Y I = 0$. Moreover, putting $Y' = \mathrm{sgrad}\, v_l$ $(l \neq m)$, we have

$$i_{[Y,Y']}\omega = -\mathcal{L}_Y(dv_l) = -d(Yv_l) = 0.$$

Hence $[Y, Y'] = 0$, and (3) follows. □

2. Summing up the local data

In the previous section we have given, for each $p \in M^1$, the neighborhood U, the constants $e_{\alpha\beta}$, the functions h_i, a_i, b_{ij}, v_i, and the basis $F_1, \dots, F_n$ of $\mathcal{F}$, as well as their numbering. From now on, to clarify the point-dependence, we shall write $U^{(p)}$, $h_i^{(p)}$, $F_i^{(p)}$, etc. instead. We assume that each neighborhood $U^{(p)}$ is taken to be a small distance ball centered at p so that it is convex.

Take a point $p_0 \in M^1$ and fix it. Let $M^{1,0}$ be the connected component of M^1 containing p_0. Let $p \in M^{1,0}$, and let $\gamma(t)$ $(0 \le t \le 1)$ be a curve in $M^{1,0}$ such that $\gamma(0) = p_0$, $\gamma(1) = p$. Along the curve γ there is a unique numbering of $\{D_i\}$ so that $t \mapsto (D_i)_{\gamma(t)}$ is continuous.

Since the relations $i \preceq j$ are locally stable, we have the following

LEMMA 2.1. *The relation $\preceq$ on $\{1, \dots, n\}$ is constant along the curve γ. In particular, the partially ordered set $\mathcal{A}$ is constantly defined along γ.*

We put

$$F_\alpha^{(q)}(\lambda) = \sum_{i \in \alpha} (-\lambda)^{i-s(\alpha)} F_i^{(q)}.$$

As is easily seen,

$$F_\alpha^{(q)}(\lambda) = |u_\alpha^{(q)}| \sum_{j\in\alpha} \prod_{\substack{k\in\alpha \\ k\neq j}} (h_k^{(q)} - \lambda) \cdot (V_j^2 + (IV_j)^2)$$

$$+ |u_\alpha^{(q)}| \sum_{\beta \in \mathfrak{n}(\alpha)} \frac{\prod_{l\in\alpha}(h_l^{(q)} + e_{\alpha\beta}^{(q)}) - \prod_{l\in\alpha}(h_l^{(q)} - \lambda)}{e_{\alpha\beta}^{(q)} + \lambda} \sum_{\gamma \succeq \beta} \sum_{j \in \gamma} (V_j^2 + (IV_j)^2).$$

PROPOSITION 2.2. *There are constants $c_\alpha \neq 0$ and d_α such that*

$$F_\alpha^{(p)}(c_\alpha \lambda - d_\alpha) = \left(c_\alpha^{|\alpha|-1} \prod_{\beta \preceq \alpha} c_\beta^{|\beta|} \right) F_\alpha^{(p_0)}(\lambda).$$

If $\{h_i^{(p)}\}$ and $e_{\alpha\beta}^{(p)}$ are suitably chosen, then those constants become

$$c_\alpha = 1, \quad d_\alpha = 0 \qquad (\textit{any } \alpha).$$

PROOF. Let I_t be the connected component of the set of t' such that $\gamma(t') \in U^{(\gamma(t))}$ containing the point t. Suppose that $I_{t_1} \cap I_{t_2} \neq \emptyset$. Then, by Proposition 1.11 there are constants $\widetilde{c}_\alpha (\neq 0)$ and $\widetilde{d}_\alpha$ such that

$$h_i^{(\gamma(t_1))} = \widetilde{c}_\alpha h_i^{(\gamma(t_2))} - \widetilde{d}_\alpha, \quad e_{\alpha\beta}^{(\gamma(t_1))} = \widetilde{c}_\alpha e_{\alpha\beta}^{(\gamma(t_2))} + \widetilde{d}_\alpha$$

on $U^{(\gamma(t_1))} \cap U^{(\gamma(t_2))}$. This implies that

$$F_\alpha^{(\gamma(t_1))}(\widetilde{c}_\alpha \lambda - \widetilde{d}_\alpha) = \left(\widetilde{c}_\alpha^{|\alpha|-1} \prod_{\beta \preceq \alpha} \widetilde{c}_\beta^{|\beta|} \right) F_\alpha^{(\gamma(t_2))}(\lambda).$$

Taking a finite number of points on γ and iterating this argument successively, we obtain the former half of the proposition.

Now, putting

$$\widetilde{h}_i^{(p)} = c_\alpha^{-1}(h_i^{(p)} + d_\alpha), \quad \widetilde{e}_{\alpha\beta}^{(p)} = c_\alpha^{-1}(e_{\alpha\beta}^{(p)} - d_\alpha),$$

and denoting by $\widetilde{F}_\alpha^{(p)}(\lambda)$ the corresponding polynomial, we have

$$\widetilde{F}_\alpha^{(p)}(\lambda) = \left(c_\alpha^{|\alpha|-1} \prod_{\beta \preceq \alpha} c_\beta^{|\beta|} \right)^{-1} F_\alpha^{(p)}(c_\alpha \lambda - d_\alpha) = F_\alpha^{(p_0)}(\lambda).$$

□

Now, let us suppose that $p = p_0$, i.e., γ is a loop. Let ν be the permutation of the indices $1, \dots, n$ defined by

$$(D_i)_{\gamma(1)} = (D_{\nu(i)})_{\gamma(0)}.$$

PROPOSITION 2.3. *ν is the identity.*

PROOF. Since ν preserves the relation $\preceq$, it induces an automorphism of the partially ordered set. We also denote it by ν. Then, by Proposition 2.2 there are constants $c_\alpha \neq 0$ and d_α such that

$$F_{\nu(\alpha)}^{(p_0)}(c_\alpha \lambda - d_\alpha) = \left(c_\alpha^{|\alpha|-1} \prod_{\beta \preceq \alpha} c_\beta^{|\beta|} \right) F_\alpha^{(p_0)}(\lambda).$$

This formula clearly indicates that $\nu(\alpha) = \alpha$ and $F_i^{(p)}$ $(i \in \alpha)$ is written as a linear combination of $F_{\nu(j)}^{(p)}$ $(i \leq j \leq t(\alpha))$. Thus $\nu(i) = i$ for every $i \in \alpha$. □

This proposition implies that the subbundles D_i are globally defined on $M^{1,0}$.

PROPOSITION 2.4. *Suppose that $\{h_i^{(q)}\}$ and $e_{\alpha\beta}^{(q)}$ $(q \in M^{1,0})$ are takcn so that $F_\alpha^{(q)}(\lambda) = F_\alpha^{(p_0)}(\lambda)$ for all $\alpha \in \mathcal{A}$. Then for any $p, q \in M^{1,0}$ such that $U^{(p)} \cap U^{(q)} \neq \emptyset$, $e_{\alpha\beta}^{(p)} = e_{\alpha\beta}^{(q)}$ and $h_i^{(p)} = h_i^{(q)}$ for any i on the intersection. Hence, there are functions $\{h_i\}$ on $M^{1,0}$ such that $h_i = h_i^{(p)}$ on $U^{(p)}$.*

PROOF. Since $F_\alpha^{(p)}(\lambda) = F_\alpha^{(q)}(\lambda)$, the proposition follows from the proof of Proposition 2.2. □

3. Structure of $M - M^1$

In the previous section we have obtained the constants $e_{\alpha\beta}$, the functions h_i, a_i, b_{ij}, f_{ij} on $M^{1,0}$ and the basis $\{F_i\}$ of $\mathcal{F}$. Also, the functions $v \in \mathcal{V}$ and the vector fields $Y \in \mathfrak{k}$ are now defined on $M^{1,0}$, and the properties described in Proposition 1.16 bolds on $M^{1,0}$. Since for each α the functions h_i $(i \in \alpha)$ take mutually distinct values at every point in $M^{1,0}$, we may assume that the numbering of $\{D_i\}$ is chosen so that

$$h_{s(\alpha)} > h_{s(\alpha)+1} > \cdots > h_{t(\alpha)} \qquad (\alpha \in \mathcal{A})$$

on $M^{1,0}$. We put

$$\begin{aligned} v_\alpha(\lambda) &= u_\alpha \prod_{i \in \alpha} (h_i - \lambda) \\ &= \sum_{m=0}^{|\alpha|-1} (-\lambda)^m v_{s(\alpha)+m} + (-\lambda)^{|\alpha|} u_\alpha. \end{aligned}$$

The main purpose of this section is to prove the following

THEOREM 3.1. *$M - M^1$ is equal with a locally finite union of closed, totally geodesic, complex hypersurfaces L. In particular M^1 is connected and dense in M. Moreover, the functions h_i are continuously extended to the whole M and the subbundles D_i are smoothly extended to M^0, and they possess the following properties:*

(1) *$h_i + e_{\alpha\beta}$ ($i \in \alpha$, $\alpha \prec \beta$) are everywhere nonzero on M;*
(2) *$h_{s(\alpha)} > h_{s(\alpha)+1} > \cdots > h_{t(\alpha)}$ on M^0 ($\alpha \in \mathcal{A}$);*
(3) *h_i are C^∞ functions on M^0;*
(4) *$\mathcal{S}_m(h_i\,;\, i \in \alpha)$ ($1 \le m \le |\alpha|$) are C^∞ functions on M;*
(5) *$D_\alpha = \sum_{i\in\alpha} D_i$ ($\alpha \in \mathcal{A}$) are extended to the whole M as C^∞ subbundles of TM;*
(6) *The vector fields $Y \in \mathfrak{k}$ are globally defined and of C^∞ on M;*
(7) *For each hypersurface L there are $\alpha \in \mathcal{A}$ and $c \in \boldsymbol{R}$ such that a connected component of the set of zeros of $\operatorname{sgrad} v_\alpha(c) \in \mathfrak{k}$ coincides with L, and $v_\alpha(c)$ vanishes on L.*

We shall call $\{h_i\}$ and $\{e_{\alpha\beta}\}$ *the fundamental functions* and *the conjunction constants* of the Kähler-Liouville manifold $(M, \mathcal{F})$ (of type (A)) respectively. Note that if $\{h'_i\}$ and $\{e'_{\alpha\beta}\}$ are other choice, then there are constants $c_\alpha \neq 0$ and d_α such that $e'_{\alpha\beta} = c_\alpha e_{\alpha\beta} + d_\alpha$ and for $i \in \alpha$,

$$h'_i = \begin{cases} c_\alpha h_i - d_\alpha & (c_\alpha > 0) \\ c_\alpha h_{t(\alpha)+s(\alpha)-i} - d_\alpha & (c_\alpha < 0) \end{cases}$$

The following theorem is an immediate consequence of the theorem above and Proposition 1.16.

THEOREM 3.2. *The geodesic flow of a Kähler-Liouville manifold of type (A) is integrable with respect to the first integrals in $\mathcal{F}$ and $\mathfrak{k}$.*

We begin the proof of Theorem 3.1 with a characterization of the functions v_i in terms of F_j.

LEMMA 3.3. *Let $i \in \alpha$. If $i \neq s(\alpha)$, the function v_i is equal with a linear combination of $\operatorname{tr} F_{i-1}, \dots, \operatorname{tr} F_{t(\alpha)}$ and $\operatorname{tr} F_j$ ($j \in \beta$, $\beta \prec \alpha$). If $i = s(\alpha)$ and α is not a maximal element, then v_i is equal with a linear combination of $\operatorname{tr} F_j$ ($j \in \beta$, $\beta \preceq \alpha$) and $\operatorname{tr} F_{t(\gamma)}$, where γ is any element of $\mathfrak{n}(\alpha)$. Finally, if $i = s(\alpha)$ and α is a maximal element, then $v_i u_\alpha^{|\alpha|-1}$ is a polynomial of $\det F_{t(\alpha)-1}$ and $\operatorname{tr} F_j$ ($j \subset \beta$, $\beta \preceq \alpha$).*

The proof is straightforward. The lemma above implies that there is a C^∞ function on M expressed by the traces of F_j that coincides with v_i on $M^{1,0}$, unless $i = s(\alpha)$ and α is maximal. Though the expression is not unique in general, we take one and fix it. We denote the extended function by the same symbol v_i. Note that every u_β ($\beta \in \mathcal{A}$) is a linear combination of those functions. Similarly, if $i = s(\alpha)$ and α is maximal, then there is a C^∞ function on M that coincides with $v_i u_\alpha^{|\alpha|-1}$ on $M^{1,0}$.

Let $M^{0,0}$ be the connected component of M^0 that contains $M^{1,0}$. For technical reason we shall first show that $M^{0,0} \cap M^1$ is connected and dense in $M^{0,0}$, and that the functions h_i are smoothly extended to $M^{0,0}$. Let $c(t)$ be a geodesic such that

$$p = c(0) \in M^{1,0}, \quad c([0, t_0)) \subset M^{1,0}, \quad q = c(t_0) \in M^{0,0} - M^{1,0}.$$

Since $q \in M^0$, every simultaneous eigenspace of $\{F_q^e \mid F \in \mathcal{F}\}$ is (complex) one-dimensional. Hence there is a neighborhood U of q, and there are subbundles D_i of TM on U such that each D_i coincides with that on $M^{1,0}$ on the connected component of $U \cap M^{1,0}$ containing a curve segment of the form $c((t_0 - \epsilon, t_0))$, $\epsilon > 0$.

Since the polynomial $v_\alpha(\lambda)$ has $|\alpha|$ real roots at each point in $M^{1,0}$, so does at q if $u_\alpha(q) \neq 0$. In this case, denoting those roots by $h_i(q)$ $(i \in \alpha)$, $h_{s(\alpha)}(q) \geq \cdots \geq h_{t(\alpha)}(q)$, we have the continuous extension of h_i up to q.

LEMMA 3.4. (1) *$u_\alpha(q) \neq 0$ for any α.*
(2) $h_{s(\alpha)}(q) > \cdots > h_{t(\alpha)}(q)$.

PROOF. If $u_\alpha(q) = 0$ for some α, then we have $F_{t(\alpha)} = 0$ at q, contradicting $q \in M^0$. Hence $u_\alpha(q) \neq 0$ for every α, and the functions h_i are well-defined at q. Since the eigenvalues of the endomorphism F_i^e at each point in $M^{1,0}$ are given by f_{ij} described in Proposition 1.15, so are at q by continuity. Therefore, if $h_i(q) = h_{i+1}(q)$ for some $i, i+1 \in \alpha$, then one can easily see that D_i and D_{i+1} cannot be separated by means of the eigenvalues, which contradicts the facts that every simultaneous eigenspace is one-dimensional. □

By virtue of the lemma above, we see that the polynomial $v_\alpha(\lambda)$ has $|\alpha|$ distinct real roots on U, if U is taken small enough. Denoting those roots again by h_i $(i \in \alpha)$, $h_{s(\alpha)} > \cdots > h_{t(\alpha)}$, we obtain the smooth extension of the functions h_i to $M^{1,0} \cup U$.

LEMMA 3.5. *There is some h_i such that $dh_i = 0$ at q. In this case we also have:*
(1) *The hessian* $\operatorname{Hess} h_i$ *of h_i at q is given by*

$$\operatorname{Hess} h_i(X, Y) = ag([X]_{D_i}, [Y]_{D_i}), \qquad X, Y \in T_qM,$$

where $a \in \boldsymbol{R}$, $a \neq 0$, and $[X]_{D_i}$ denotes the D_i-component of X;
(2) *$\dot{c}(t_0)$ is not orthogonal to D_i.*

PROOF. Put $U' = M^{1,0} \cap U$, and let $\{\widetilde{a}_i\}$ be functions around q given in Proposition 1.1. Then we have

$$d \log \widetilde{a}_i \equiv -d \log \left(|u_\alpha| \prod_{\substack{j \in \alpha \\ j \neq i}} |h_j - h_i| \right) \qquad \text{mod } D_i$$

on U'. Clearly, it is also valid on the closure of U' in U by continuity. Hence, if every derivatives dh_i does not vanish at q, then it follows that $q \in M^1$, a contradiction. Thus there is some i such that $dh_i = 0$ at q.

Put $b = h_i(q)$ and suppose $i \in \alpha$. Then $dv_\alpha(b) = 0$ and

$$\operatorname{Hess} v_\alpha(b) = \widetilde{a} \operatorname{Hess} h_i$$

at q, where $\widetilde{a}$ is a non-zero constant. Put $Y = \operatorname{sgrad} v_\alpha(b)$. Then we have

$$\operatorname{Hess} v_\alpha(b)(X, Z) = g(\nabla_X Y, IZ).$$

Moreover, since Y is an infinitesimal automorphism of the Kähler manifold M on U', we have $\nabla_{IX} Y = I \nabla_X Y$. Hence $\operatorname{Hess} v_\alpha(b)$ is a hermitian form at q. Since $(\operatorname{Hess} h_i)(X, Z) = 0$ at q if X or Z is orthogonal to D_i, it follows that

$$(\operatorname{Hess} h_i)(X, Z) = (\operatorname{Hess} h_i)([X]_{D_i}, [Z]_{D_i}) = ag([X]_{D_i}, [Z]_{D_i})$$

at q for some constant a.

Now, we show $(\mathrm{Hess}\, v_\alpha(b))(\dot c(t_0), \cdot) \neq 0$, which will prove that $a \neq 0$, and (2). Since Y is a Jacobi field along $c(t)$ $(0 \leq t < t_0)$, it satisfies the equation of Jacobi field up to $q = c(t_0)$ by continuity. Hence, Y_q being 0, we have $\nabla_{\dot c(t_0)} Y \neq 0$ at q. From this it follows that

$$(\mathrm{Hess}\, v_\alpha(b))(\dot c(t_0), \cdot) \neq 0.$$

□

LEMMA 3.6. *There is a constant* $\epsilon > 0$ *such that* $c((t_0, t_0 + \epsilon)) \subset M^{1,0}$.

PROOF. Let S be the subspace of T_qM spanned by $\dot c(t_0)$ and $I\dot c(t_0)$, and let B be the image of the ϵ-ball $\{V \in S \mid |V| < \epsilon\}$ in S by the exponential mapping $\mathrm{Exp}_q : T_qM \to M$. Then, it follows from the previous lemma that $dh_i \neq 0$ at q' for every i and $q' \in B - \{q\}$, provided ϵ small enough. It also follows from the proof of the previous lemma that if $q' \in U$ lies on the boundary of U', then some dh_i vanishes at q'. Hence we have $B - \{q\} \subset M^{1,0}$. □

PROPOSITION 3.7. $M^{0,0} \cap M^1 = M^{1,0}$, *and it is dense in* $M^{0,0}$. *Also, the functions* h_i *and the subbundles* D_i *extend smoothly to* $M^{0,0}$ *and satisfies*

$$h_{s(\alpha)} > \cdots > h_{t(\alpha)} \quad (\alpha \in \mathcal{A}), \quad dh_i|_{D_j} = 0 \quad \text{if } i \neq j.$$

Moreover, D_i *are the simultaneous eigenspaces of the endomorphisms* $\{F^e \mid F \in \mathcal{F}\}$ *at each point in* $M^{0,0}$.

PROOF. Let N be the closure of $M^{1,0}$ in $M^{0,0}$, and assume that $N \neq M^{0,0}$. Let $q \in N$ be a boundary point of N, and let U be an open distance ball centered at q, which is small enough so that it is convex and contained in $M^{0,0}$. Let p_0 and q_0 be two points in U such that $p_0 \in M^{1,0}$ and $q_0 \notin N$. Let $c(t)$ $(0 \leq t \leq T)$ be the minimal geodesic from p_0 to q_0, and let $t = t_1$ be the earliest time when $c(t)$ meets the boundary of N.

Applying the argument above to the geodesic $c(t)$ and the point $p_1 = c(t_1)$, we see that there is $\epsilon > 0$ such that $c((t_1, t_1 + \epsilon)) \subset M^{1,0}$. Iterating this procedure successively, we obtain a sequence of times $0 = t_0 < t_1 < t_2 < \cdots < T$ such that

$$c(t_k) \notin M^{1,0}, \quad c((t_{k-1}, t_k)) \subset M^{1,0} \qquad (k \geq 1).$$

We claim that the number of such t_k is finite. In fact, suppose that it is not the case, and put $t_\infty = \lim_{k\to\infty} t_k \leq T$. Then, as seen above, any vector fields of the form

$$Y = \mathrm{sgrad}\, v_\alpha(b) \qquad (\alpha \in \mathcal{A},\ b \in \boldsymbol{R})$$

are Jacobi fields along $c(t)$ $(0 \leq t \leq t_\infty)$, and among them there is Y_k that vanish at $c(t_k)$ for every $k \geq 1$. Let $\tilde{\mathfrak{k}}$ be the vector space of Jacobi fields spanned by such Y. Since $\tilde{\mathfrak{k}}$ coincides with $\mathfrak{k}$ on $c([0,T]) \cap M^{1,0}$, it follows that it is n-dimensional, and satisfies

$$g(Y, \nabla_{\dot c} Y') = g(\nabla_{\dot c} Y, Y') \qquad Y, Y' \in \tilde{\mathfrak{k}}.$$

From this property one can easily conclude that the set of points t such that some non-zero $Y \in \tilde{\mathfrak{k}}$ vanishes at $c(t)$ is discrete. On the other hand, choosing a subsequence if necessary, we obtain $Y_\infty \in \tilde{\mathfrak{k}} - \{0\}$ as a limit of (constant multiples of) $\{Y_k\}$. Since Y_∞ vanishes at $c(t_\infty)$, it is a contradiction. Hence there are only finitely many t_k; $t_1, \dots, t_l$.

Now, again by Lemma 3.6 we see that $c((t_l, T)) \subset M^{1,0}$. However, this contradicts the fact that $q_0 = c(T) \notin N$. Thus we conclude that $N = M^{0,0}$, and $M^{0,0} \cap M^1 = M^{1,0}$. The remaining part is obvious. □

Next, we shall prove that M^0 is connected and dense in M. Let q be a boundary point of $M^{0,0}$, and assume that $u_\alpha(q) \neq 0$ for every α. Then in the same way as above the functions h_i are continuously extended to $M^{0,0} \cup \{q\}$. Also, T_qM is decomposed to the simultaneous eigenspaces of the endomorphisms $\{F^e \mid F \in \mathcal{F}\}$. Clearly those subspaces are uniquely extended as C^∞ subbundles around q so that they are sums of simultaneous eigenspaces at each point.

LEMMA 3.8. *Let $q \in M^s \cap \overline{M^{0,0}}$, and assume that $u_\alpha(q) \neq 0$ for any α. Let D be one of the simultaneous eigenspaces of $\{F^e \mid F \in \mathcal{F}\}$ at q, and let* $\dim D = m$. *Then there is $\alpha \in \mathcal{A}$ and its subset α' consisting of successive numbers $i, \dots, i+m-1$ such that the extended subbundle (also denoted by D) is equal with $\sum_{j\in\alpha'} D_j$ on $M^{0,0}$ near q. Moreover, there is a constant c such that:*

1. *$h_i(q) = \cdots = h_{i+m-1}(q) = c, \qquad h_j(q) \neq c \quad (j \in \alpha - \alpha')$;*
2. *The functions $\mathcal{S}_k(h_j\,;\, j \in \alpha')$ on $M^{0,0}$ can be smoothly extended around q $(1 \le k \le m)$;*
3. *$dv_\alpha(c) = 0$ at q if $m \geq 2$;*
4. *The hessian of $v_\alpha(c)$ vanishes at q if $m \geq 3$.*

Also, there exists at least one simultaneous eigenspace D with $\dim D \geq 2$.

PROOF. Let U be a neighborhood of q where the subbundle D is defined. Since D is a sum of simultaneous eigenspaces at each point, it is a sum of D_j on $U \cap M^{0,0}$. Let us consider the endomorphisms $F^e_{t(\alpha)}$. It is $|u_\alpha|$ times the identity on $\sum_{\beta \succeq \alpha} D_\beta$, and 0 on the orthogonal complement. Therefore the subbundles D_α are continuously extended to q so that $\sum_{\beta\succeq\alpha} D_\beta$ is still the eigenspace of $F^e_{t(\alpha)}$ corresponding to the eigenvalue $|u_\alpha(q)|$. Since D_α is a sum of simultaneous eigenspaces at q, it is consequently extended on U as the subbundle of TM. Clearly $D \subset D_\alpha$ for some α. Also, the endomorphism F^e_j at q is a constant multiple of the identity on D_α if $j \notin \alpha$, and the eigenvalues of F^e_j $(j \in \alpha)$ on D_α are

$$|u_\alpha(q)|\mathcal{S}_{t(\alpha)-j}(h_l(q)\,;\, l \in \alpha - \{k\}) \qquad (k \in \alpha)$$

at q. This implies that there is a subset α' of α consisting of successive numbers $i, \dots, i+m-1$ such that

$$D = \sum_{j\in\alpha'} D_j \quad \text{on } U \cap M^{0,0}$$
$$h_i(q) = \cdots = h_{i+m-1}(q) = c$$
$$h_j(q) \neq c \qquad \text{for } j \in \alpha - \alpha'.$$

To prove (2) we note that $\sum_{j\in\alpha} h_j$ is extended as the C^∞ function around q, because $v_{t(\alpha)}$ and u_α are of C^∞, and $u_\alpha(q) \neq 0$. Hence the symmetric functions of the eigenvalues of the endomorphisms

$$|u_\alpha|^{-1}\left(\left(\sum_{j\in\alpha} h_j\right) F^e_{t(\alpha)} - F^e_{t(\alpha)-1}\right)$$

on D, which are

$$\mathcal{S}_l(h_j;\ j \in \alpha'),$$

on $M^{0,0}$, are C^∞ functions around q. This proves (2).

To prove the assertions (3) and (4), assume $m \geq 2$, and let $G(\lambda)$ be the endomorphism of D_α defined to be the restriction of

$$F^e_{t(\alpha)-1} - \left(\sum_{j\in\alpha} h_j(q) - \lambda\right) F^e_{t(\alpha)}$$

to D_α. Then $\det G(\lambda)$ is a C^∞ function on U. Since $G(c) = 0$ on D at q, the order of the zero q of the function $\det G(c)$ is not less than m. On the other hand, we have

$$\begin{aligned}\det G(\lambda) &= |u_\alpha|^{|\alpha|} \prod_{j\in\alpha}\left(\sum_{k\in\alpha} h_k - \sum_{k\in\alpha} h_k(q) - (h_j - \lambda)\right)\\ &= \epsilon^{|\alpha|} \sum_{l=0}^{|\alpha|} (-1)^l u_\alpha^l \mathcal{S}_l(h_j - \lambda\,;\, j \in \alpha)\left(v_{t(\alpha)} - u_\alpha \sum_{k\in\alpha} h_k(q)\right)^{|\alpha|-l}\end{aligned}$$

on $U \cap \overline{M^{0,0}}$, where ϵ is the sign of u_α. Note that $u_\alpha \mathcal{S}_l(h_j - \lambda\,;\, j \in \alpha)$ is expressed as a linear combination of v_j ($j \in \alpha$) and u_α. Hence the last formula described above also expresses a C^∞ function on U that coincides with $\det G(\lambda)$ on $U \cap \overline{M^{0,0}}$. In particular those two functions coincides at q up to infinite order. Since $d(\det G(c)) = 0$ at q, and since

$$\mathcal{S}_{|\alpha|-1}(h_j - c\,;\, j \in \alpha) = v_{t(\alpha)} - u_\alpha \sum_{k\in\alpha} h_k(q) = 0$$

at q, it follows that

$$d(u_\alpha \mathcal{S}_{|\alpha|}(h_j - c\,;\, j \in \alpha)) = 0$$

at q. Moreover, if $m \geq 3$, considering the function $(d/d\lambda)\det G(\lambda)|_{\lambda=c}$, we also have

$$d(u_\alpha \mathcal{S}_{|\alpha|-1}(h_j - c\,;\, j \in \alpha)) = 0$$

at q. Then observing the function $\det G(c)$ again, we see that the hessian of the function

$$u_\alpha \mathcal{S}_{|\alpha|}(h_j - c\,;\, j \in \alpha) = v_\alpha(c)$$

vanishes at q.

Finally, assume that every simultaneous eigenspace is of dimension one. Then as is easily seen, the $|\alpha|$ functions h_i ($i \in \alpha$) take mutually different values at q for any α. Hence the functions $a_{ij} = a_i b_{ij}$ can be continuously extended to q, which implies the linear independence of $F_1, \ldots F_n$ at q. □

PROPOSITION 3.9. *M^0 is connected and dense in M.*

PROOF. Assume that M^0 is not connected, and let $p, p' \in M^0$ such that $p \in M^{1,0}$ and p' lies in another component. Let $c(t)$ $(0 \leq t \leq t_0)$ be a geodesic from p to p'. A slight modification of the geodesic c and the point p' enables us to assume that D_i-component of $\dot{c}(0) \in T_pM$ is not zero for any i. Let t_1 be the time such that $q = c(t_1) \in M^s$ and $c(t) \in M^{0,0}$ for any $t \in [0, t_1)$.

We first claim that every (extended) function u_α ($\alpha \in \mathcal{A}$) does not vanish at q. In fact, assume that $u_\alpha(q) = 0$. Since

$$\operatorname{tr} F_{t(\alpha)} = |u_\alpha| \sum_{\beta \succeq \alpha} |\beta|$$

and since $F_{t(\alpha)}$ is positive semi-definite on $M^{0,0}$, it follows that $F_{t(\alpha)} = 0$ at q. Let ζ_t be the geodesic flow and $\pi : T^*M \to M$ the bundle projection. Then $c(t) = \pi(\zeta_t \lambda_0)$, where $\lambda_0 \in T_p^*M$ is given by

$$\lambda_0(X) = g(\dot{c}(0), X) \qquad X \in T_pM,$$

and we have

$$F_{t(\alpha)}(\lambda_0) = F_{t(\alpha)}(\zeta_{t_1}\lambda_0) = 0.$$

This implies that $\dot{c}(0)$ is orthogonal to D_j for any $j \in \beta$, $\beta \succeq \alpha$, a contradiction. Hence $u_\alpha(q) \neq 0$ for every α.

Therefore, as stated above, the functions h_i is extended up to q, and Lemma 3.8 is applicable. Let D be a simultaneous eigenspace of the endomorphisms $\{F^e \mid F \in \mathcal{F}\}$ at q, and let α and α' be as in Lemma 3.8. Suppose $m = \dim D \geq 2$, and put $h_j(q) = a$ ($j \in \alpha'$). Then $Y = \operatorname{sgrad} v_\alpha(a)$ is the Jacobi field along the geodesic $c|_{[0,t_1]}$, and vanishes at $q = c(t_1)$. Hence $\nabla_{\dot{c}(t_1)} Y \neq 0$, which implies that the hessian of $v_\alpha(a)$ at q does not vanish. Thus we have $m = 2$ by Lemma 3.8.

Put $\mathcal{F}' = \{F \in \mathcal{F} \mid F_q = 0\}$, and let $\dim \mathcal{F}' = k$. For each 2-dimensional simultaneous eigenspace D, put

$$H_D = F_\alpha(a) - \sum_{\beta \in \mathfrak{n}(\alpha)} \epsilon_\beta (e_{\alpha\beta} + a)^{-1} F_{t(\beta)},$$

where ϵ_β is the sign of $\prod_{l \in \alpha}(h_l + e_{\alpha\beta})$. As is easily seen, the elements H_D form a basis of $\mathcal{F}'$. For $F \in \mathcal{F}$, let X_F be the vector field on T^*M defined by

$$i_{X_F} d\theta = -dF,$$

where θ is the canonical 1-form on the cotangent bundle. Clearly, the vector field X_F ($F \in \mathcal{F}'$) on T^*M is tangent to the fibre T_q^*M at each point $\lambda \in T_q^*M$. Thus we have the vector field $X_F|_{T_q^*M}$ on the vector space T_q^*M whose coefficients are hermitian forms. Let S_q^*M be the sphere of unit covectors at q, and put

$$\Lambda = \{\lambda \in S_q^*M \mid X_F|_{T_q^*M} = 0 \text{ at } \lambda \text{ for some } F \in \mathcal{F}' - \{0\}\}.$$

LEMMA 3.10. (1) $(X_F)_{\zeta_{t_1}\lambda_0} \neq 0$ *for any* $F \in \mathcal{F} - \{0\}$.

(2) *For each 2-dimensional simultaneous eigenspace* D, *there is a unit vector* $V_D \in D$ *such that* Λ *is given by*

$$\Lambda = \{\lambda \in S_q^*M \mid \lambda(V_D) = \lambda(IV_D) = 0 \text{ for some } D\}.$$

In particular the complement of Λ *in* S_q^*M *is connected and dense in* S_q^*M.

PROOF. Describing
$$F_p = \sum_j b_j(V_j^2 + (IV_j)^2),$$
we have
$$\pi_*((X_F)_{\lambda_0}) = 2\sum_j b_j(g(V_j, \dot{c}(0))V_j + g(IV_j, \dot{c}(0))IV_j).$$
Since the right-hand side does not vanish because of the assumption on $\dot{c}(0)$, it follows that
$$(X_F)_{\zeta_{t_1}\lambda_0} = \zeta_{t_1 *}((X_F)_{\lambda_0}) \neq 0.$$

As is easily seen, the endomorphism H_D^e on the orthogonal complement $D^\perp$ of D vanishes up to order 2 at q. Hence, taking an orthonormal frame $\widetilde{V}_i, I\widetilde{V}_i$ $(i = 1, 2)$ of D around q, we see that $X_{H_D}|_{T_q^*M}$ is of the form
$$X_{H_D}|_{T_q^*M} = f_1\widetilde{V}_1^* + f_2\widetilde{V}_2^* + f_3(I\widetilde{V}_1)^* + f_4(I\widetilde{V}_2)^*,$$
where $\widetilde{V}_1^*, \ldots, (I\widetilde{V}_2)^*$ are covectors (constant vector fields on T_q^*M) that vanish on $D^\perp$, and dual to $\widetilde{V}_1, \ldots, I\widetilde{V}_2$; also, f_i are linear combinations of the hermitian forms
$$\widetilde{V}_i^2 + (I\widetilde{V}_i)^2 \quad (i = 1, 2), \quad \widetilde{V}_1\widetilde{V}_2 + (I\widetilde{V}_1)(I\widetilde{V}_2), \quad \widetilde{V}_1 I\widetilde{V}_2 - \widetilde{V}_2 I\widetilde{V}_1.$$
Then, since
$$0 = X_{H_D}F_{t(\alpha)} = |u(\alpha)(q)| X_{H_D} \sum_{i=1}^{2} (\widetilde{V}_i^2 + (I\widetilde{V}_i)^2),$$
we can choose $\widetilde{V}_1$ and $\widetilde{V}_2$ so that
$$\begin{aligned} aX_{H_D} &= (\widetilde{V}_1^2 + (I\widetilde{V}_1)^2)\widetilde{V}_2^* - (\widetilde{V}_1\widetilde{V}_2 + (I\widetilde{V}_1)(I\widetilde{V}_2))\widetilde{V}_1^* \\ &\quad + (\widetilde{V}_1 I\widetilde{V}_2 - \widetilde{V}_2 I\widetilde{V}_1)(I\widetilde{V}_1)^*, \end{aligned}$$
where a is a non-zero constant, and the covectors $\widetilde{V}_i^*, (I\widetilde{V}_i)^*$ are identified with the constant vector fields on T_q^*M. Hence, the zero set of X_{H_D} on T_q^*M is the vector subspace defined by $\widetilde{V}_1 = I\widetilde{V}_1 = 0$. By putting $V_D = \widetilde{V}_1$, (2) follows. □

Let Λ_1 be the set of points $\lambda \in S_q^*M$ such that
$$\dim\{(X_F)_\lambda \mid F \in \mathcal{F}\} < n.$$

LEMMA 3.11. (1) *$S_q^*M - \Lambda_1$ is connected and dense in S_q^*M.*

(2) *For $\lambda \in S_q^*M - \Lambda_1$ the set of $t \in \mathbf{R}$ such that $\pi(\zeta_t\lambda) \notin M^0$ is discrete. In particular there is a constant $\epsilon > 0$ such that $\pi(\zeta_t\lambda) \in M^{0,0}$ for t satisfying $|t| < \epsilon, t \neq 0$.*

PROOF. We define V_D for each 1-dimensional simultaneous eigenspace D as a unit vector in D. Let Λ_2 be the set of $\lambda \in S_q^*M$ such that
$$\lambda(V_D) = \lambda(IV_D) = 0$$
for some simultaneous eigenspace D (of dimension 1 or 2). Then the complement of Λ_2 in S_q^*M is still connected and dense in S_q^*M. We prove that
$$S_q^*M - \Lambda_2 \subset S_q^*M - \Lambda_1,$$
which will indicate (1).

Let $\lambda \in S^*_q M - \Lambda_2$. Since $\lambda \notin \Lambda$, the previous lemma implies that $(X_F)_\lambda$ ($F \in \mathcal{F}'$) form a k-dimensional subspace of $T_\lambda(T^*_q M)$. On the other hand, we know that $\{F_q \mid F \in \mathcal{F}\}$ is $(n-k)$-dimensional, and spanned by

$$\widetilde{V}_1^2 + \widetilde{V}_2^2 + (I\widetilde{V}_1)^2 + (I\widetilde{V}_2)^2 \quad (D, 2\text{-dimensional}),$$
$$(V_D)^2 + (IV_D)^2 \quad (D, 1\text{-dimensional}),$$

where $\widetilde{V}_1, \ldots, I\widetilde{V}_2$ is an orthonormal basis of D. Hence, it follows that

$$\{\pi_*((X_F)_\lambda) \mid F \in \mathcal{F}\}$$

is $(n-k)$-dimensional subspace of T_qM, provided $\lambda \notin \Lambda_2$. From these two facts we see that

$$\dim\{(X_F)_\lambda \mid F \in \mathcal{F}\} = n \qquad (\lambda \in S^*_q M - \Lambda_2).$$

Hence $\lambda \in S^*_q M - \Lambda_1$.

Now, let $\lambda \in S^*_q M - \Lambda_1$, and put $Z_F(t) = \pi_*((X_F)_{\zeta_t \lambda})$. Then, $Z_F(t)$ are Jacobi fields along the geodesic $c_1(t) = \pi(\zeta_t \lambda)$, and satisfy

$$g(Z_F, \nabla_{\dot{c}_1} Z_{\widetilde{F}}) = g(\nabla_{\dot{c}_1} Z_F, Z_{\widetilde{F}}) \qquad (F, \widetilde{F} \in \mathcal{F}).$$

This implies that the set of t at which $Z_F(t) = 0$ for some $F \in \mathcal{F} - \{0\}$ is discrete. Since $F_{c_1(t)} = 0$ implies $Z_F(t) = 0$ for each t, it follows that $\pi(\zeta_t \lambda) \in M^0$ except discrete t's. □

We now continue the proof of Proposition 3.9. The lemmas above imply that there are only finite number of t on the interval $(0, t_0)$ such that $c(t) \notin M^0$. Let $t_1, \ldots, t_l$ ($t_1 < \cdots < t_l < t_0$) be those points. Then the previous lemma also implies that $c((t_1, t_2)) \subset M^{0,0}$. Since Lemmas 3.10 and 3.11 are still valid for the point $c(t_2)$, it also follows that $c((t_2, t_3)) \subset M^{0,0}$. Iterating this procedure successively, we consequently see that $p' = c(t_0) \in M^{0,0}$, which shows the connectedness of M^0.

The denseness of M^0 in M is now clear, because any geodesic $c(t)$ emanating from the point p meets M^s only at discrete values of t, if every D_i-component of $\dot{c}(0)$ is nonzero. This completes the proof of Proposition 3.9. □

We have proved that M^0 and M^1 are connected and dense in M, and u_α does not vanish everywhere on M^0 for any α. Now we shall show that $u_\alpha \neq 0$ everywhere on M. Assume that $u_\alpha(q) = 0$ for some $\alpha \in \mathcal{A}$ and $q \in M^s$. Then $du_\alpha = 0$ at q, because u_α is either non-positive or non-negative on M. Since the functions u_β are globally defined and smooth on M, the vector fields $\operatorname{sgrad} u_\beta$ are globally defined infinitesimal automorphisms of the Kähler manifold M. Therefore the connected component L of the set of zeros of $\operatorname{sgrad} u_\alpha$ containing q is a totally geodesic complex submanifold of M (see **[Kob2]**, for instance). Also, the tangent space of L at q coincides with the kernel of the hessian of u_α at q.

Let U be a small neighborhood of q, and let $p \in U \cap M^1$ that does not belong to any such submanifolds L corresponding to the vector fields $\operatorname{sgrad} u_\beta$ vanishing at q. Let $c(t)$ be the minimal geodesic joining q and p; $c(0) = q$, $c(t_0) = p$. Fix α such that $u_\alpha(q) = 0$, and put

$$v = v_{t(\alpha)} - \left(\sum_{i \in \alpha} h_i(p)\right) u_\alpha.$$

Note that the function v is well-defined and smooth on the whole M.

LEMMA 3.12. $\operatorname{Hess} v(\dot{c}(0), \dot{c}(0)) = 0$.

PROOF. Taking U small enough, we may assume that $u_\gamma(c(t)) \neq 0$ for any $\gamma \in \mathcal{A}$. Hence, as we have seen before, there is an open and dense subset J of the interval $(0, t_0]$ such that $c(t) \in M^1$ for $t \in J$. Since $F_{t(\alpha)}$ is semi-definite everywhere, and since $\operatorname{tr} F_{t(\alpha)}$ is a non-zero multiple of u_α, we have $F_{t(\alpha)} = 0$ at q. Hence $F_{t(\alpha)}(\zeta_t\lambda) = 0$ for any $t \in \boldsymbol{R}$ ($c(t) = \pi(\zeta_t\lambda)$), and this implies that

$$\dot{c}(t) \in \sum_{\gamma \not\leq \alpha} D_\gamma \qquad (t \in J).$$

Hence $\sum_{i \in \alpha} h_i(c(t))$ is constant on each connected component of J. So, by continuity we have

$$v(c(t)) = u_\alpha(c(t)) \left(\sum_{i \in \alpha} h_i(c(t)) - \sum_{i \in \alpha} h_i(p) \right) = 0$$

for all $t \in [0, t_0]$, and it therefore follows that $\operatorname{Hess} v(\dot{c}(0), \dot{c}(0)) = 0$. □

It is clear that a slight modification of the initial vector $\dot{c}(0) \in T_qM$ does not affect the conclusion of Lemma 3.12. Thus we have $\operatorname{Hess} v = 0$ at q, which contradicts the fact that $\operatorname{sgrad} v$ is the non-trivial Killing vector field on M. Hence it has been shown that u_α does not vanish everywhere on M for every α.

We now prove the remaining part of Theorem 3.1. We have already shown that the subbundles D_i are well-defined and of C^∞ on M^0. Also, (1), (2), and (3) have been verified. Since the functions u_α are everywhere non-zero, (4) and (6) also follows. Let q be a point in $M - M^1$. If $c(t)$ is a geodesic passing through q and a point in M^1, then the set of $t \in \boldsymbol{R}$ such that $\{Y_{c(t)} \mid Y \in \mathfrak{t}\}$ is not n-dimensional is discrete, as we have already seen. Hence Lemma 3.5 and the argument in the proof of Proposition 3.9 is applicable. Thus if $q \in M^0 - M^1$, there are $i (\in \alpha)$ and $a, b \in \boldsymbol{R}$, $b \neq 0$ such that $d(v_\alpha(a)) = 0$ at q and

$$\operatorname{Hess} v_\alpha(a)(X, Y) = bg([x]_{D_i}, [Y]_{D_i}), \qquad X, Y \in T_qM.$$

Therefore the connected component L of the zeros of the infinitesimal automorphism $\operatorname{sgrad} v_\alpha(a)$ passing through q is the complex submanifold of codimension one.

Now let $q \in M^e$. Then it follows that every simultaneous eigenspace of $\{F_q^e \mid F \in \mathcal{F}\}$ is of dimension one or two, and they are contained in some D_α. Hence (5) follows. Let D be a simultaneous eigenspace of dimension two, and let α, $\alpha' = \{i, i+1\}$ and $c \in \boldsymbol{R}$ be as in Lemma 3.8. From the proof of Lemma 3.8, it easily follows that $d(v_\alpha(c)) = 0$ and $dv \neq 0$ at q, where

$$v = u_\alpha S_{|\alpha|-1}(h_j - c;\, j \in \alpha)$$

This implies that the exterior derivative of the function $h_i + h_{i+1}$, which is of C^∞ around q, does not vanish at q. Since $d((h_i - c)(h_{i+1} - c)) = 0$ at q, it follows that $\operatorname{grad} v \in D$, and the orthogonal complement of D is contained in the kernel of $\operatorname{Hess} v_\alpha(c)$ at q. Since $dv_\alpha(c)(\operatorname{sgrad} v) = 0$ everywhere, it therefore follows that the connected component L of the set of zeros of $\operatorname{sgrad} v_\alpha(c)$ passing through q is also $(n-1)$-dimensional.

Thus $M - M^1$ is equal with the union of such hypersurfaces L. The local finiteness is clear. This complete the proof of Theorem 3.1.

Also, we have just proved the following

PROPOSITION 3.13. *Let $q \in M^s$. Then the simultaneous eigenspaces D of the linear endomorphisms $\{F_q^e \mid F \in \mathcal{F}\}$ are of dimension one or two. They are smoothly extended to a neighborhood U of q as the subbundles of TM in the following way: If $\dim D = 1$, then $D = D_i$ on $U \cap M^0$ for some i; if $\dim D = 2$, then there is $\alpha \in \mathcal{A}$ and $i, i+1 \in \alpha$ such that $D = D_i + D_{i+1}$ on $U \cap M^0$. In the first case, the function h_i is smooth around q, and if $i \in \alpha$,*

$$h_j(q) \neq h_i(q) \qquad (j \in \alpha,\ j \neq i).$$

In the second case,

$$h_i(q) = h_{i+1}(q), \quad h_j(q) \neq h_i(q) \qquad (j \in \alpha,\ j \neq i, i+1),$$

and the functions h_i and h_{i+1} are not differentiable at q. Moreover, $h_i + h_{i+1}$ and $h_i h_{i+1}$ are smooth functions around q such that $d(h_i + h_{i+1}) \neq 0$ and $d((h_i - h_i(q))(h_{i+1} - h_i(q))) = 0$ at q.

4. Torus action and the invariant hypersurfaces

In the rest of the paper we shall assume that the Kähler-Liouville manifold M is *compact*. Let $\mathfrak{k}$ be as before, and put

$$\mathfrak{g} = \mathfrak{k} + I\mathfrak{k},$$

which is a commutative Lie algebra of infinitesimal holomorphic transformations of M. Let K and G be the Lie transformation group of M generated by $\mathfrak{k}$ and $\mathfrak{g}$ respectively. Note that $\mathfrak{g}$ is naturally regarded as a complex Lie algebra. Accordingly, G is regarded as a complex Lie group so that the action $G \times M \to M$ is holomorphic. In this section we shall investigate the properties of the action of those groups and the hypersurfaces contained in $M - M^1$.

We first prove the following

PROPOSITION 4.1. (1) *G preserves M^1 and each hypersurface $L \subset M - M^1$.*
(2) *The action of G on M^1 is simply transitive.*

PROOF. To prove (1), recall that L is a connected component of the set of zeros of $\operatorname{sgrad} v \in \mathfrak{k}$ for some $v \in \mathcal{V}$. Therefore, every $Y \in \mathfrak{k}$ is tangent to L. Since L is a complex submanifold, any $Z \in \mathfrak{g}$ is also tangent to L. Thus G preserves L. Since M^1 is the complement of the union of such hypersurfaces, it is also preserved by G. To prove (2), note that $\{Y_p \mid Y \in \mathfrak{g}\}$ is real $2n$-dimensional for every point $p \in M^1$. Hence the G-action on M^1 is transitive. Suppose $gp = p$ for some $g \in G$ and $p \in M^1$. Then $gq = q$ for every $q \in M^1$, because G is abelian. Hence g should be the identity transformation of M by continuity. □

THEOREM 4.2. *The Lie group K is isomorphic to*

$$U(1)^n = U(1) \times \cdots \times U(1) \quad (n \text{ times}),$$

where $U(1)$ is the group of unit complex numbers. Also, G is isomorphic to $(\boldsymbol{C}^\times)^n$ as complex Lie group, where $\boldsymbol{C}^\times$ is the multiplicative group of non-zero complex numbers.

To prove this theorem we need several lemmas. Let L be a complex hypersurface contained in $M - M^1$. As observed before, there is $v_\alpha(c) \in \mathcal{V}$ such that L coincides with a connected component of the set of zeros of $\operatorname{sgrad} v_\alpha(c)$, and $v_\alpha(c)$ vanishes on L. In this case we shall call $v_\alpha(c)$ *the function that determines* L.

LEMMA 4.3. *Let L and $v_\alpha(c)$ be as above. Then the vector field* $\operatorname{sgrad} v_\alpha(c)$ *generates a circle action on M.*

PROOF. Put $Y = \operatorname{sgrad} v_\alpha(c)$, and let $\operatorname{ad} Y$ be the linear endomorphism of T_pM $(p \in L)$ given by the formula

$$(\operatorname{ad} Y)(X) = [Y, X](= -\nabla_X Y), \qquad X \in T_pM.$$

Clearly, $\operatorname{ad} Y$ preserves the normal space N_pL. Since N_pL is complex 1-dimensional and Y is the infinitesimal isometry, there is $a \in \boldsymbol{R}$ such that

$$(\operatorname{ad} Y)(X) = aIX \quad \text{for any} \quad X \in N_pL.$$

We claim that a is independent of $p \in L$. In fact, let $Z \in T_pL$, and let X be a unit normal vector field along L. Then

$$-\nabla_Z(\nabla_X Y) = (Za)IX + aI\nabla_Z X.$$

Since Y is a Killing vector field, we have

$$\nabla_Z(\nabla_X Y) = \nabla_{\nabla_Z X} Y + \frac{1}{2}(R(Z,X)Y - R(Y,Z)X + R(X,Y)Z).$$

Hence $Za = 0$, which shows that a is constant on L.

The lemma is now clear, because the linear isotropy action of the 1-parameter group ϕ_t generated by Y has the least period $2\pi/|a|$ at every point $p \in L$, and so is ϕ_t itself via the exponential mapping $\operatorname{Exp}\colon NL \to M$. □

LEMMA 4.4. *Fix $\alpha \in \mathcal{A}$, and take a constant b so that $h_i + b > 0$ on M for any $i \in \alpha$. Put*

$$v = u_\alpha \prod_{i \in \alpha} (h_i + b) \in \mathcal{V},$$

and let $p \in M$ be a point where $|v|$ attains the maximum. Then the functions h_j are smooth around p, and $dh_j = 0$ at p for any j such that $j \in \beta$, $\beta \preceq \alpha$.

PROOF. If $p \in M^0$, then the assertion is clear, because $dh_i(D_j) = 0$ for any i, j such that $i \neq j$. Similarly, if there is no 2-dimensional simultaneous eigenspace of $\{F_p^e \mid F \in \mathcal{F}\}$ contained in $\sum_{\beta \preceq \alpha} D_\beta$, the assertion is also clear. Now, assume that $p \in M^s$ and there are $\beta \preceq \alpha$ and $i, i+1 \in \beta$ such that $h_i(p) = h_{i+1}(p)$. Let D be the subbundle of TM defined on a neighborhood U of p that coincides with $D_i + D_{i+1}$ on $M^0 \cap U$. Then, putting

$$X = \operatorname{grad}(h_i + h_{i+1}),$$

we have $X \in D$ and $X \neq 0$ by virtue of Proposition 3.13. Since

$$d((h_i - h_i(p))(h_{i+1} - h_i(p))) = 0$$

at p, it therefore follows that

$$d((h_i + b)(h_{i+1} + b))(X) \neq 0 \qquad \text{at } p.$$

However, this implies that $dv(X) \neq 0$ at p, a contradiction. □

PROOF OF THEOREM 4.2. Fix $\alpha \in \mathcal{A}$, and let $p \in M$ be as in Lemma 4.4. Clearly the function $v_\alpha(h_i(p))$ $(i \in \alpha)$ determines a hypersurface L_i contained in $M - M^1$ that pass through p. Hence the vector fields

$$Y_i = \operatorname{sgrad} v_\alpha(h_i(p)) \qquad (i \in \alpha)$$

generate circle actions on M by Lemma 4.3. Executing the same procedure for all α we thus obtain n vector fields $Y_i \in \mathfrak{k}$ each of which generates a circle action on M. As is easily seen, those vector fields form a basis of $\mathfrak{k}$. Hence K is compact, and is isomorphic to $U(1)^n$.

Now, fix a point $q \in M^1$, and let $\phi_i : \mathfrak{g} \to \boldsymbol{R}$ be the mapping defined by

$$\phi_i(Z) = v_\alpha(h_i(p))((\exp Z)q).$$

Put $\Phi = (\phi_i) : \mathfrak{g} \to \boldsymbol{R}^n$. Then we have $\Phi(Z + Y) = \Phi(Z)$ for any $Y \in \mathfrak{k}$ and $Z \in \mathfrak{g}$, and

$$\frac{\partial}{\partial t_j}\phi_i(\sum_k t_k IY_k) = -g(Y_i, Y_j).$$

From this formula it easily follows that the inner product of two vectors

$$\Phi(\sum_k t_k IY_k) - \Phi(\sum_k s_k IY_k) \quad \text{and} \quad t - s = (t_k - s_k)$$

in $\boldsymbol{R}^n$ is negative, provided $t \neq s$. Hence $\Phi|_{I\mathfrak{k}}$ is injective.

These facts indicate that the kernel of the homomorphism $\exp : \mathfrak{g} \to G$ is equal to that of $\exp : \mathfrak{k} \to K$. This proves the latter half of the theorem. □

Our next goal is to determine all the hypersurfaces contained in $M - M^1$. For this purpose we need deeper information on the fundamental functions h_i. Let L be a hypersurface in $M - M^1$ determined by the function $v_\alpha(c) \in \mathcal{V}$. The following lemma is easy.

LEMMA 4.5. *Let $p \in L$. Then there are neighborhoods W and U of p in L and M respectively, a neighborhood V of 0 in $\boldsymbol{C} = \{(z)\}$, and a holomorphic diffeomorphism $\phi : W \times V \to U$ such that*

(1) *$\phi(q, 0) = q$ for any $q \in W$,*

(2) *$\phi_* z\frac{\partial}{\partial z} = (2a)^{-1}(-IY - \sqrt{-1}Y)$,*

where $Y = \operatorname{sgrad} v_\alpha(c)$ and a is the eigenvalue of $I \circ \operatorname{ad} Y$ on NL.

Note that the hessian of $v_\alpha(c)$ on the normal bundle NL is equal to a times the metric g, a being the constant in Lemma 4.5. Hence there is a neighborhood U of L such that $v_\alpha(c) > 0$ (resp. < 0) on $U - L$ if $a > 0$ (resp. $a < 0$).

LEMMA 4.6. *$v_\alpha(c) > 0$ (resp. < 0) on $M - L$ if $a > 0$ (resp. $a < 0$).*

PROOF. Suppose $a > 0$. Let ψ_t be the one-parameter group of transformations generated by $-\operatorname{grad} v_\alpha(c)$. Then the previous lemma implies that $\psi_t(q)$ converges to a point in L as $t \to \infty$ for any $q \in U$, provided U ($\supset L$) small enough. Now, fix $q_0 \in U$, and let $q \in M^1$ be an arbitrary point. Then there is $g \in G$ such that $q = gq_0$. Since $\psi_t(q) = g\psi_t(q_0)$ and $gL = L$, it follows that $\psi_t(q) \in U$ for sufficiently large t. Hence we have

$$v_\alpha(c)(q) > v_\alpha(c)(\psi_t(q)) > 0.$$

Let $b > 0$ be the minimal value of the function $v_\alpha(c)$ on the boundary of U, and let q' be a point in $M - (M^1 \cup U)$. Then one can take a point $q \in M^1 - U$ arbitrary near q'. Since $v_\alpha(c)(q) > b$, it follows that $v_\alpha(c)(q') \geq b$, proving the lemma. The case $a < 0$ is similar. □

LEMMA 4.7. *Let $\alpha \in \mathcal{A}$, and let $i, i+1 \in \alpha$. Then*

$$\min h_i \geq \max h_{i+1},$$

where the minimum and the maximum are taken on M.

PROOF. Suppose that h_i takes its minimal value c at $p \in M$. In this case we have $h_{i-1}(p) > c$ if $i - 1 \in \alpha$. In fact, if $i - 1 \in \alpha$ and $h_{i-1}(p) = c$, then c is also the minimal value of h_{i-1}. Hence $d(h_{i-1} + h_i) = 0$ at p, contradicting Proposition 3.13. Then, there are two cases: (1) $h_{i+1}(p) < c$, or (2) $h_i(p) = h_{i+1}(p) = c$. If the case (1) occurs, then the function h_i is smooth around p. Hence the function $v_\alpha(c)$ determines a hypersurface $L \subset M - M^1$. Since $h_{i+1} \neq c$ on $M - L$ by the previous lemma, we have $\max h_{i+1} \leq c$.

Now, suppose that the case (2) occurs. Then by virtue of Proposition 3.13 the function $v_\alpha(c)$ again determines a hypersurface $L \subset M - M^1$. Proposition 3.13 also implies that there is a point $q \in L$ near p such that $h_{i+1}(q) < c$ and $h_i(q) = c$. Hence we again conclude that $\max h_{i+1} \leq c$. □

PROPOSITION 4.8. *A function in $\mathcal{V}$ of the form $v_\alpha(c)$ determines a hypersurface L in $M - M^1$ if and only if c is the maximal or minimal value of h_i for some $i \in \alpha$.*

PROOF. The "if" part has been indicated in the proof of the previous proposition. Now, suppose that a function $v_\alpha(c)$ determines a hypersurface L in $M - M^1$, and let $p \in L$. Then there is $i \in \alpha$ such that $h_i(p) = c$. Since $h_i \neq c$ on $M - L$, c should be the maximal value or the minimal value of h_i. □

COROLLARY 4.9. *Suppose that the function h_i is smooth around a point $p \in M$, and $dh_i = 0$ at p. Then h_i takes its maximum or minimum at p.*

PROOF. The assumption implies that the function $v_\alpha(h_i(p))$ determines a hypersurface in $M - M^1$ ($i \in \alpha$). Thus the corollary follows from Proposition 4.8. □

We now prove the following theorem.

THEOREM 4.10. *For any $\alpha \in \mathcal{A}$ and $i, i+1 \in \alpha$,*

$$\min h_i = \max h_{i+1}.$$

The following corollary is an immediate consequence of Theorem 4.10 and Proposition 4.8.

COROLLARY 4.11. *Put*

$$c_{\alpha,0} = \max h_{s(\alpha)}, \qquad c_{\alpha,|\alpha|} = \min h_{t(\alpha)},$$
$$c_{\alpha,\nu} = \min h_{s(\alpha)+\nu-1} = \max h_{s(\alpha)+\nu} \quad (1 \le \nu \le |\alpha| - 1),$$

and let $L_{\alpha,\nu}$ be the hypersurface in $M - M^1$ determined by $v_\alpha(c_{\alpha,\nu})$. Then the hypersurfaces $L_{\alpha,\nu}$ are mutually distinct, and the set

$$\{L_{\alpha,\nu} \mid \alpha \in \mathcal{A},\ 0 \le \nu \le |\alpha|\}$$

coincides with the set of all closed hypersurfaces contained in $M - M^1$.

Let us recall that the fundamental functions h_i $(i \in \alpha)$ and the conjunction constants $e_{\alpha\beta}$ $(\alpha \prec \beta)$ may be replaced with

$$k_\alpha h_i - l_\alpha \quad (k_\alpha > 0) \quad \text{or} \quad k_\alpha h_{t(\alpha)+s(\alpha)-i} - l_\alpha \quad (k_\alpha < 0)$$

and $k_\alpha e_{\alpha\beta} + l_\alpha$ respectively, where $k_\alpha \neq 0$ and l_α are constants. Hence it is always possible to choose h_i and $e_{\alpha\beta}$ so that

$$1 = c_{\alpha,0} > c_{\alpha,1} > \cdots > c_{\alpha,|\alpha|} = 0. \tag{4.1}$$

In this case we also have

$$e_{\alpha\beta} > 0 \quad \text{or} \quad e_{\alpha\beta} < -1. \tag{4.2}$$

Under this condition the only possible alternative choice of h_i $(i \in \alpha)$ and $e_{\alpha\beta}$ are given by

$$h'_i = 1 - h_{t(\alpha)+s(\alpha)-i} \quad (i \in \alpha), \qquad e'_{\alpha\beta} = -1 - e_{\alpha\beta}.$$

In the rest of the paper we shall always assume that *the fundamental functions $\{h_i\}$ and the conjunction constants $\{e_{\alpha\beta}\}$ are chosen so that the condition (4.1) is satisfied.* Also, we shall call $\{c_{\alpha,\nu}\}$ *the fundamental constants.*

PROOF OF THEOREM 4.10. Assume that $i, i+1 \in \alpha$ and

$$\min h_i = c_2 > c_3 = \max h_{i+1},$$

and put $c_1 = \max h_i$, $c_4 = \min h_{i+1}$. Let L_μ be the hypersurface in $M - M^1$ determined by $v_\alpha(c_\mu)$, and put

$$Y_\mu = \operatorname{sgrad} v_\alpha(c_\mu) \qquad (\mu = 1, \ldots, 4).$$

Also, put

$$b_j = \begin{cases} \max h_j & (j < i) \\ \min h_j & (j > i+1) \end{cases}$$

for $j \in \alpha$, $j \neq i, i+1$.

LEMMA 4.12. *There are four points $p_{13}, p_{14}, p_{23}, p_{24}$ such that*

(1) $p_{\mu\nu} \in L_\mu \cap L_\nu$ $\quad (\mu = 1, 2,\ \nu = 3, 4)$,

(2) *h_j is smooth and $dh_j = 0$ at the four points for every $\beta \preceq \alpha$ and every $j \in \beta$,*

(3) *$h_j = b_j$ at the four points for every $j \in \alpha$, $j \neq i, i+1$.*

PROOF. First we show that $L_\mu \cap L_\nu \neq \emptyset$ ($\mu = 1, 2,\ \nu = 3, 4$). Let b the maximal value of the function h_i on the hypersurface L_3, and suppose $h_i(q) = b$, $q \in L_3$. If $h_{i-1}(q) = b$, then $b = c_1$. Otherwise, the function h_i is smooth and $dh_i = 0$ at q. Hence $b = c_1$ or c_2 by Corollary 4.9. Since the case $b = c_2$ contradicts the choice of q, we have $b = c_1$. Thus $L_1 \cap L_3 \neq \emptyset$. The other cases are similar.

Now, choose a constant d such that $c_2 > d > c_3$, and let $p_{\mu\nu} \in L_\mu \cap L_\nu$ be a point where the function $|v_\alpha(d)|$, restricted to $L_\mu \cap L_\nu$, takes the maximal value ($\mu = 1, 2,\ \nu = 3, 4$). Then the similar argument as the proof of Lemma 4.4 clearly indicates that the conditions (2) and (3) are satisfied. □

Let us consider the following identity (the decomposition to linear fractions):

$$\begin{aligned}\frac{v_\alpha(\lambda)}{\prod_{j\in\alpha'}(b_j-\lambda)\prod_{\mu=1}^{3}(c_\mu-\lambda)} &= \sum_{j\in\alpha'}\frac{1}{b_j-\lambda}\frac{v_\alpha(b_j)}{\prod_{\substack{k\in\alpha'\\ k\neq j}}(b_k-b_j)\prod_{\mu=1}^{3}(c_\mu-b_j)}\\ &\quad+\sum_{\mu=1}^{3}\frac{1}{c_\mu-\lambda}\frac{v_\alpha(c_\mu)}{\prod_{j\in\alpha'}(b_j-c_\mu)\prod_{\substack{1\le\nu\le3\\ \nu\neq\mu}}(c_\nu-c_\mu)},\end{aligned}\tag{4.3}$$

where $\alpha' = \alpha - \{i, i+1\}$. Multiplying both sides by $-\lambda$ and taking the limit $\lambda \to \infty$, we obtain the identity:

$$\begin{aligned}u_\alpha &= \sum_{j\in\alpha'}\frac{v_\alpha(b_j)}{\prod_{\substack{k\in\alpha'\\ k\neq j}}(b_k-b_j)\prod_{\mu=1}^{3}(c_\mu-b_j)}\\ &\quad+\sum_{\mu=1}^{3}\frac{v_\alpha(c_\mu)}{\prod_{j\in\alpha'}(b_j-c_\mu)\prod_{\substack{1\le\nu\le3\\ \nu\neq\mu}}(c_\nu-c_\mu)}.\end{aligned}$$

This formula gives the linear dependence of the skew-gradient vector fields Y_1, Y_2, Y_3, $\operatorname{sgrad} v_\alpha(b_j)$ ($j \in \alpha'$), and $\operatorname{sgrad} u_\alpha$. Let d_μ be the positive number so that the least period of the one-parameter group generated by Y_μ is $2\pi/d_\mu$ ($\mu = 1, \ldots, 4$). We now consider the linear isotropy action of the one-parameter groups generated by those vector fields at the point p_{13} (Note that those vector fields vanish at p_{13}). There we have the decomposition

$$T_{p_{13}}M = N_{p_{13}}L_1 \oplus N_{p_{13}}L_3 \oplus T_{p_{13}}(L_1 \cap L_3).$$

Clearly, the linear isotropy action of the one-parameter groups generated by Y_3, $\operatorname{sgrad} u_\alpha$, and $\operatorname{sgrad} v_\alpha(b_j)$ ($j \in \alpha'$) are trivial on $N_{p_{13}}L_1$. Hence the linear endomorphism

$$\exp\left(\frac{tY_2}{\prod_{j\in\alpha'}(b_j-c_2)\prod_{\substack{1\le\mu\le3\\ \mu\neq2}}(c_\mu-c_2)}\right)\exp\left(\frac{tY_1}{\prod_{j\in\alpha'}(b_j-c_1)\prod_{\substack{1\le\mu\le3\\ \mu\neq1}}(c_\mu-c_1)}\right)\tag{4.4}$$

of $N_{p_{13}}L_1$ is trivial for all $t \in \boldsymbol{R}$. Substituting

$$t = 2\pi d_2^{-1}(c_1-c_2)(c_3-c_2)\prod_{j\in\alpha'}(b_j-c_2)$$

in (4.4), we conclude that

$$2\pi d_2^{-1}(c_1-c_2)(c_3-c_2)\prod_{j\in\alpha'}(b_j-c_2) = m\cdot 2\pi d_1^{-1}(c_2-c_1)(c_3-c_1)\prod_{j\in\alpha'}(b_j-c_1)\tag{4.5}$$

for some $m \in \mathbf{Z}$. The similar argument at the point p_{23} gives the formula

$$2\pi d_1^{-1}(c_2 - c_1)(c_3 - c_1) \prod_{j\in\alpha'} (b_j - c_1) = m' \cdot 2\pi d_2^{-1}(c_1 - c_2)(c_3 - c_2) \prod_{j\in\alpha'} (b_j - c_2)$$

for some $m' \in \mathbf{Z}$. Therefore $m = \pm 1$. Replacing c_3 by c_4 in the formula (4.3), and considering the linear isotropy actions at p_{14} and p_{24}, we also obtain the formula

$$(4.6)\quad 2\pi d_2^{-1}(c_1 - c_2)(c_4 - c_2) \prod_{j\in\alpha'} (b_j - c_2) = \pm 2\pi d_1^{-1}(c_2 - c_1)(c_4 - c_1) \prod_{j\in\alpha'} (b_j - c_1).$$

From (4.5) and (4.6) we have

$$(c_3 - c_2)(c_4 - c_1) = \pm(c_4 - c_2)(c_3 - c_1),$$

which contradicts the inequality; $c_1 > c_2 > c_3 > c_4$. This completes the proof of Theorem 4.10. □

In the rest of the section we shall observe the detail of the action of G on M. In particular we shall show that M is a toric variety. Let $c_{\alpha,\nu}$ and $L_{\alpha,\nu}$ be as in Corollary 4.11, and put

$$\mathcal{J} = \{(\alpha, \nu) \mid \alpha \in \mathcal{A},\ 0 \le \nu \le |\alpha|\}.$$

Let $d_{\alpha,\nu}$ be the non-zero constant such that

$$\operatorname{Hess} v_\alpha(c_{\alpha,\nu})(X, X) = d_{\alpha,\nu} g(X, X) \quad (X \in NL_{\alpha,\nu}),$$

and put

$$Y_{\alpha,\nu} = d_{\alpha,\nu}^{-1} \operatorname{sgrad} v_\alpha(c_{\alpha,\nu}) \in \mathfrak{k}.$$

Also, let $\mathcal{I}$ be the set of sections of the mapping $\mathcal{J} \to \mathcal{A}$ $((\alpha,\nu) \mapsto \alpha)$, that is, $\iota \in \mathcal{I}$ is an assignment of an index $\iota(\alpha)$ $(0 \le \iota(\alpha) \le |\alpha|)$ to each $\alpha \in \mathcal{A}$. Put

$$\mathcal{J}(\iota) = \{(\alpha, \nu) \in \mathcal{J} \mid \nu \ne \iota(\alpha)\} \qquad (\iota \in \mathcal{I}).$$

LEMMA 4.13. (1) *For any* α, $\cap_{\nu=0}^{|\alpha|} L_{\alpha,\nu} = \emptyset$.
(2) *For any* $\iota \in \mathcal{I}$, $\cap_{(\alpha,\nu)\in\mathcal{J}(\iota)} L_{\alpha,\nu} \ne \emptyset$.

PROOF. (1) Let $p \in \cap_{\nu=0}^{|\alpha|} L_{\alpha,\nu}$. First, since $p \in L_{\alpha,0}$, it follows that $h_{s(\alpha)}(p) = c_{\alpha,0}$. Next, the condition that $p \in L_{\alpha,1}$ implies that $h_{s(\alpha)+1}(p) = c_{\alpha,1}$, and so on. Consequently, we have $h_{s(\alpha)+\nu}(p) = c_{\alpha,\nu}$ $(0 \le \nu \le |\alpha| - 1)$ from the condition that $p \in \cap_{\nu=0}^{|\alpha|-1} L_{\alpha,\nu}$. However, since $p \in L_{\alpha,|\alpha|}$, we also have $h_{t(\alpha)}(p) = c_{\alpha,|\alpha|}$, a contradiction.

(2) Put

$$v = \prod_{\alpha\in\mathcal{A}} \prod_{i\in\alpha} (h_i - c_{\alpha,\iota(\alpha)}),$$

and let $p \in M$ be a point where the function $|v|$ takes the maximum. Then, as in the proof of Lemma 4.4, we see that $p \in M^0$ and $dh_i = 0$ at p for every i. Hence

$$\{h_i(p) \mid i \in \alpha\} = \{c_{\alpha,\nu} \mid 0 \le \nu \le |\alpha|,\ \nu \ne \iota(\alpha)\}.$$

This indicates that $p \in L_{\alpha,\nu}$ for every $(\alpha, \nu) \in \mathcal{J}(\iota)$. □

Let U be the neighborhood of $L_{\alpha,\nu}$ given in the proof of Lemma 4.6, and put

$$U_{\alpha,\nu} = \cup_{t\in\mathbf{R}} \psi_t(U),$$

where ψ_t is the one-parameter group generated by $-IY_{\alpha,\nu}$. It is clear from Lemma 4.6 that $U_{\alpha,\nu}$ is G-invariant, and $\psi_t(q)$ converges to a point in $L_{\alpha,\nu}$ as $t \to -\infty$ for any $q \in U_{\alpha,\nu}$. Define the mapping $\rho_{\alpha,\nu} : U_{\alpha,\nu} \to L_{\alpha,\nu}$ by

$$\rho_{\alpha,\nu}(q) = \lim_{t\to-\infty} \psi_t(q).$$

PROPOSITION 4.14. (1) $U_{\alpha,\nu} = M - \bigcap_{\substack{0\le\mu\le|\alpha| \\ \mu\neq\nu}} L_{\alpha,\mu}$.

(2) $\rho_{\alpha,\nu} : U_{\alpha,\nu} \to L_{\alpha,\nu}$ *is the holomorphic fibre bundle with typical fibre* $\boldsymbol{C}$. *Also,* $\rho_{\alpha,\nu} : U_{\alpha,\nu} - L_{\alpha,\nu} \to L_{\alpha,\nu}$ *is the principal* $\boldsymbol{C}^\times$*-bundle.*

PROOF. Put

$$S = \bigcap_{\substack{0\le\mu\le|\alpha| \\ \mu\neq\nu}} L_{\alpha,\mu}.$$

Since S is G-invariant, and $S \cap L_{\alpha,\nu} = \emptyset$, it follows that $U_{\alpha,\nu} \subset M - S$. Now, we shall show the reversed inclusion. Put

$$v = \prod_{i\in\alpha}(h_i - c_{\alpha,\nu}) = u_\alpha^{-1} v_\alpha(c_{\alpha,\nu}).$$

Then, the function $|v|$ is positive on $M - L_{\alpha,\nu}$ and takes its maximal value

$$\prod_{\substack{0\le\mu\le|\alpha| \\ \mu\neq\nu}} |c_{\alpha,\mu} - c_{\alpha,\nu}|$$

on S. Also, we have

$$\frac{d}{dt}|v(\psi_t(q))| = |\mathrm{grad}\, v|^2 \cdot \frac{d_{\alpha,\nu}^{-1} v_\alpha(c_{\alpha,\nu})}{|v|} > 0$$

for every $q \in U_{\alpha,\nu} - L_{\alpha,\nu}$ and $t \in \boldsymbol{R}$.

Let $q \in M - S$. Let $p_j = \psi_{t_j}(q)$ $(0 \ge t_j \to -\infty)$ be a converging sequence, and let $p \in M$ be its limit point. Since $|v(p)| < |v(q)|$, it follows that $p \in M - S$. Also, we have $(d|v|)_p = 0$. As is easily seen, the set of critical points of the function $|v|$ is equal to $L_{\alpha,\nu} \cup S$. Hence it follows that $p \in L_{\alpha,\nu}$. This implies that $\psi_{-t}(q) \in U$ for sufficiently large t. Thus $q \in U_{\alpha,\nu}$, completing the proof of (1). (2) is the immediate consequence of Lemma 4.5. □

PROPOSITION 4.15. *Let* $\mathcal{J}_0$ *be a subset of* $\mathcal{J}$ *such that* $\bigcap_{(\alpha,\nu)\in\mathcal{J}_0} L_{\alpha,\nu} \neq \emptyset$. *Let* ρ_0 *be the composition of all* $\rho_{\alpha,\nu}$, $(\alpha,\nu) \in \mathcal{J}_0$, *and put*

$$S_0 = \bigcup_{\alpha\in\mathcal{A}} \bigcap_{\substack{0\le\nu\le|\alpha| \\ (\alpha,\nu)\notin\mathcal{J}_0}} L_{\alpha,\nu}.$$

Then the mapping $\rho_0 : M - S_0 \to \bigcap_{(\alpha,\nu)\in\mathcal{J}_0} L_{\alpha,\nu}$ *is well-defined, and is a fibre bundle with typical fibre* $\boldsymbol{C}^k$, *where* $k = \#\mathcal{J}_0$. *In particular,* $\bigcap_{(\alpha,\nu)\in\mathcal{J}_0} L_{\alpha,\nu}$ *is connected.*

PROOF. We shall prove the proposition by induction on k. Let $\mathcal{J}_0$, ρ_0, and S_0 be as in the statement. Suppose that $(\beta,\mu) \notin \mathcal{J}_0$, and put

$$\mathcal{J}_1 = \mathcal{J}_0 \cup \{(\beta,\mu)\}.$$

We assume that $\cap_{(\alpha,\nu)\in\mathcal{J}_1} L_{\alpha,\nu} \neq \emptyset$. ρ_1 and S_1 are similarly defined. We then have the following commutative diagram:

$$\begin{array}{ccc} M - S_1 & \xrightarrow{\rho_0} & \cap_{(\alpha,\nu)\in\mathcal{J}_0} L_{\alpha,\nu} - \cap_{\substack{0\le\nu\le|\beta| \\ \nu\neq\mu}} L_{\beta,\nu} \\ \rho_{\beta,\mu}\big\downarrow & & \big\downarrow\rho_{\beta,\mu} \\ L_{\beta,\mu} - S_0 & \xrightarrow[\rho_0]{} & \cap_{(\alpha,\nu)\in\mathcal{J}_1} L_{\alpha,\nu} \end{array}$$

Hence $\rho_1 = \rho_0 \circ \rho_{\beta,\mu}$ is well-defined. From the induction assumption and the previous proposition, the rows and the columns in the diagram are fibre bundles with fibre $\boldsymbol{C}^k$ and $\boldsymbol{C}$ respectively. Let $q \in M - S_1$, and let $p = \rho_1(q)$. Then, there is a neighborhood U of p in $\cap_{(\alpha,\nu)\in\mathcal{J}_1} L_{\alpha,\nu}$ such that $\rho_{\beta,\mu}^{-1}(U)$ and $\rho_0^{-1}(U)$ are isomorphic to $U \times \boldsymbol{C}$ and $U \times \boldsymbol{C}^k$ respectively. Moreover, for each $r \in \rho_0^{-1}(p)$ the mapping

$$\rho_0 : \rho_{\beta,\mu}^{-1}(r) \to \rho_{\beta,\mu}^{-1}(p)$$

is an isomorphism, because it commutes with the $\boldsymbol{C}^\times$-action generated by $Y_{\beta,\mu}$ and $IY_{\beta,\mu}$. Therefore the mapping

$$\rho_1^{-1}(U) \to U \times \boldsymbol{C} \times \boldsymbol{C}^k$$

defined by ρ_0 and $\rho_{\beta,\mu}$ is isomorphic, and it gives the local triviality of ρ_1. □

The following corollary is an immediate consequence of Proposition 4.15.

COROLLARY 4.16. *Let $\iota \in \mathcal{I}$, and fix a point $p_0 \in M^1$. Then there is a G-equivariant holomorphic isomorphism from $M - \cup_{\alpha\in A} L_{\alpha,\iota(\alpha)}$ to*

$$\boldsymbol{C}^n = \{(z^{(\iota)}_{\alpha,\nu};\ (\alpha,\nu) \in \mathcal{J}(\iota))\}$$

such that p_0 corresponds to the point given by $z^{(\iota)}_{\alpha,\nu} = 1$ for every $(\alpha,\nu) \in \mathcal{J}(\iota)$. Here the G-action on $\boldsymbol{C}^n$ is given by

$$\exp\left(\sum_{(\beta,\mu)\in J(\iota)} (-t_{\beta,\mu} IY_{\beta,\mu} + s_{\beta,\mu} Y_{\beta,\mu})\right)(z^{(\iota)}_{\alpha,\nu}) = \left(e^{t_{\alpha,\nu}+\sqrt{-1}s_{\alpha,\nu}} z^{(\iota)}_{\alpha,\nu}\right).$$

Let Γ be the lattice in $\mathfrak{k}$ such that $2\pi\Gamma$ is equal to the kernel of the homomorphism $\exp : \mathfrak{k} \to K$ of the abelian groups. Clearly $Y_{\alpha,\nu} \in \Gamma$ for every $(\alpha,\nu) \in \mathcal{J}$.

PROPOSITION 4.17. *For any $\iota \in \mathcal{I}$, the elements $Y_{\alpha,\nu}$ $((\alpha,\nu) \in \mathcal{J}(\iota))$ form a $\boldsymbol{Z}$-basis of Γ.*

PROOF. Fix $\iota \in \mathcal{I}$. Then, by virtue of Lemma 4.13 (2) there is a point p such that $p \in L_{\alpha,\nu}$ for every $(\alpha,\nu) \in \mathcal{J}(\iota)$. Since the associated endomorphisms $\operatorname{ad} Y_{\alpha,\nu}$ of T_pM are linearly independent, it follows that $Y_{\alpha,\nu}$ $((\alpha,\nu) \in \mathcal{J}(\iota))$ form a $\boldsymbol{R}$-basis of $\mathfrak{k}$. Let Y be any element of Γ, and let

$$Y = \sum_{(\alpha,\nu)\in\mathcal{J}(\iota)} a_{\alpha,\nu} Y_{\alpha,\nu}, \quad a_{\alpha,\nu} \in \boldsymbol{R}.$$

We recall that the linear isotropy action of the one-parameter group $\exp(tY_{\alpha,\nu})$ on $N_pL_{\alpha',\nu'}$ is trivial if $(\alpha,\nu) \neq (\alpha',\nu')$, and has the least period 2π if $(\alpha,\nu) = (\alpha',\nu')$. Since $\exp(2\pi Y)$ is the identity, it thus follows that $a_{\alpha,\nu} \in \boldsymbol{Z}$ for all $(\alpha,\nu) \in \mathcal{J}(\iota)$. □

THEOREM 4.18. *M is a toric variety with respect to the action of G.*

PROOF. By virtue of Corollary 4.16, M is covered by the open sets $U_\iota = M - \cup_{\alpha\in\mathcal{A}} L_{\alpha,\iota(\alpha)}$ ($\iota \in \mathcal{I}$) each of which is holomorphically isomorphic to $\boldsymbol{C}^n$. As is easily seen, the coordinate change on $U_\iota \cap U_{\iota'}$ is given by Laurent monomials whose exponents are equal to the coefficients of the base change of Γ: $Y_{\alpha,\nu}$ ($(\alpha,\nu) \in \mathcal{J}(\iota)$) to $Y_{\alpha,\nu}$ ($(\alpha,\nu) \in \mathcal{J}(\iota')$). Hence M is an algebraic variety. Corollary 4.16 also indicates that the action of the "algebraic torus" G on M is algebraic. □

The next several propositions will give the information on the structure of the toric variety M.

PROPOSITION 4.19. *For each $\alpha \in \mathcal{A}$ the value*

$$\frac{d_{\alpha,\nu}}{\prod_{\substack{0\le\mu\le|\alpha| \\ \mu\ne\nu}}(c_{\alpha,\mu}-c_{\alpha,\nu})}$$

does not depend on ν ($0 \le \nu \le |\alpha|$).

PROOF. In the same way as the proof of Lemma 4.12 we have

$$u_\alpha = \sum_{\nu=0}^{|\alpha|} \frac{v_\alpha(c_{\alpha,\nu})}{\prod_{\substack{0\le\mu\le|\alpha| \\ \mu\ne\nu}}(c_{\alpha,\mu}-c_{\alpha,\nu})}.$$

Taking the skew gradient vector fields of both sides, we then have

$$\operatorname{sgrad} u_\alpha = \sum_{\nu=0}^{|\alpha|} \frac{d_{\alpha,\nu}}{\prod_{\substack{0\le\mu\le|\alpha| \\ \mu\ne\nu}}(c_{\alpha,\mu}-c_{\alpha,\nu})} Y_{\alpha,\nu}. \tag{4.7}$$

We claim here that $\operatorname{sgrad} u_\alpha$ is written as a linear combination of $Y_{\gamma,\nu}$ ($\gamma \prec \alpha$, $1 \le \nu \le |\gamma|$). In fact, it is clear from the definition of u_α that $\operatorname{sgrad} u_\alpha$ is written as a linear combination of $Y_{\beta,\nu}$ ($0 \le \nu \le |\beta|$), where β is the maximal element of the totally ordered set $\{\gamma \in \mathcal{A} \mid \gamma \prec \alpha\}$. Hence, by the formula (4.7) (replaced α with β) $\operatorname{sgrad} u_\alpha$ is written as a linear combination of $\operatorname{sgrad} u_\beta$ and $Y_{\beta,\nu}$ ($1 \le \nu \le |\beta|$). Thus the claim follows by induction on β.

Hence, it has been shown that each $Y_{\alpha,\nu}$ is written as a linear combination of

$$Y_{\alpha,\mu} \quad (\mu \ne \nu), \quad Y_{\gamma,\mu} \quad (\gamma \prec \alpha,\ 1 \le \mu \le |\gamma|),$$

which are part of a basis of Γ. Thus the coefficients are integers. This being true for every ν, we have

$$\frac{d_{\alpha,\nu}}{\prod_{\substack{0\le\mu\le|\alpha| \\ \mu\ne\nu}}(c_{\alpha,\mu}-c_{\alpha,\nu})} = \pm \frac{d_{\alpha,\nu'}}{\prod_{\substack{0\le\mu\le|\alpha| \\ \mu\ne\nu'}}(c_{\alpha,\mu}-c_{\alpha,\nu'})}$$

for any ν and ν'.

Note the sign of $d_{\alpha,\nu}$ is equal to the sign of the function $v_\alpha(c_{\alpha,\nu})$ on $M - L_{\alpha,\nu}$. This implies that the sign of

$$\frac{d_{\alpha,\nu}}{\prod_{\substack{0\le\mu\le|\alpha| \\ \mu\ne\nu}}(c_{\alpha,\mu}-c_{\alpha,\nu})}$$

is equal to the sign of u_α. In particular it does not depend on ν. This completes the proof of the proposition. □

We put

$$d_\alpha = \frac{d_{\alpha,\nu}}{\prod_{\substack{0\le\mu\le|\alpha|\\ \mu\ne\nu}}(c_{\alpha,\mu}-c_{\alpha,\nu})} \qquad (0\le\nu\le|\alpha|).$$

Then, $d_\alpha^{-1}\mathrm{sgrad}\, u_\alpha = \sum_{\nu=0}^{|\alpha|} Y_{\alpha,\nu} \in \Gamma$. Put

$$Z_\alpha = \sum_{\nu=0}^{|\alpha|} Y_{\alpha,\nu} \in \Gamma$$

Note that $Z_\alpha = 0$ if α is minimal. We shall call d_α ($\alpha \in \mathcal{A}$) the *scaling constants.*

For convenience we shall use two symbols $\mathfrak{p}(\alpha)$ and $\mathfrak{n}(\alpha)$: For each non-minimal $\alpha \in \mathcal{A}$, $\mathfrak{p}(\alpha)$ denotes the maximal element of the totally ordered subset $\{\gamma \in \mathcal{A} \mid \gamma \prec \alpha\}$; for each non-maximal $\alpha \in \mathcal{A}$, $\mathfrak{n}(\alpha)$ denotes the subset of $\mathcal{A}$ defined by

$$\mathfrak{n}(\alpha) = \{\gamma \in \mathcal{A} \mid \mathfrak{p}(\gamma) = \alpha\}$$

($\mathfrak{n}(\alpha)$ is identical with the one defined before). For non-minimal elements $\alpha \in \mathcal{A}$ we define constants $m_{\alpha,\nu}$ $(0 \le \nu \le |\mathfrak{p}(\alpha)|)$; putting $\beta = \mathfrak{p}(\alpha)$,

$$m_{\alpha,\nu} = \frac{d_\beta}{d_\alpha} \prod_{\substack{0\le\mu\le|\beta|\\ \mu\ne\nu}} (c_{\beta,\mu} + e_{\beta\alpha}). \tag{4.8}$$

PROPOSITION 4.20.

$$\begin{aligned} Z_\alpha &= m_{\alpha,0} Z_{\mathfrak{p}(\alpha)} + \sum_{\nu=1}^{|\mathfrak{p}(\alpha)|} (m_{\alpha,\nu} - m_{\alpha,0}) Y_{\mathfrak{p}(\alpha),\nu} \\ &= \sum_{\substack{\beta;\text{non-minimal}\\ \beta\preceq\alpha}} \left(\prod_{\beta\prec\gamma\preceq\alpha} m_{\gamma,0} \right) \sum_{\nu=1}^{|\mathfrak{p}(\beta)|} (m_{\beta,\nu} - m_{\beta,0}) Y_{\mathfrak{p}(\beta),\nu}. \end{aligned}$$

PROOF. Putting $\lambda = -e_{\beta\alpha}$ in the identity

$$\frac{v_\beta(\lambda)}{\prod_{\nu=0}^{|\beta|}(c_{\beta,\nu}-\lambda)} = \sum_{\nu=0}^{|\beta|} \frac{1}{c_{\beta,\nu}-\lambda} \frac{v_\beta(c_{\beta,\nu})}{\prod_{\mu\ne\nu}(c_{\beta,\mu}-c_{\beta,\nu})},$$

and taking the skew gradient of both sides, we have

$$\begin{aligned} Z_\alpha = &\left(\frac{d_\beta}{d_\alpha} \prod_{1\le\mu\le|\beta|} (c_{\beta,\mu} + e_{\beta\alpha}) \right) Z_\beta \\ &+ \sum_{\nu=1}^{|\beta|} \left(\frac{d_\beta}{d_\alpha} \prod_{\substack{0\le\mu\le|\beta|\\ \mu\ne\nu}} (c_{\beta,\mu} + e_{\beta\alpha}) - \frac{d_\beta}{d_\alpha} \prod_{1\le\mu\le|\beta|} (c_{\beta,\mu} + e_{\beta\alpha}) \right) Y_{\beta,\nu}. \end{aligned}$$

This proves the first equality. The second one is immediate. □

PROPOSITION 4.21. *The constants* $m_{\alpha,\nu}$ (α, *not minimal*, $0 \le \nu \le \mathfrak{p}(\alpha)$) *possess the following properties.*

(1) $m_{\alpha,\nu} - m_{\alpha,0} \in \boldsymbol{Z}$.
(2) $m_{\alpha,\nu} \in \boldsymbol{Q}$ *if* $\mathfrak{p}(\alpha)$ *is not minimal.*
(3)
$$\left(\prod_{\beta \prec \gamma \preceq \alpha} m_{\gamma,0}\right)(m_{\beta,\nu} - m_{\beta,0}) \in \boldsymbol{Z}$$
for any non-minimal β *and* α ($\succ \beta$).
(4) *Either*
$$0 < m_{\alpha,0} < \cdots < m_{\alpha,|\mathfrak{p}(\alpha)|}$$
or
$$m_{\alpha,0} > \cdots > m_{\alpha,|\mathfrak{p}(\alpha)|} > 0.$$
(5) *if* $\alpha, \alpha' \in \mathfrak{n}(\beta)$, $\alpha \neq \alpha'$, *then for any* ν, $1 \le \nu \le |\beta|$,
$$\frac{m_{\alpha,0}}{m_{\alpha,\nu}}\frac{m_{\alpha,|\beta|} - m_{\alpha,\nu}}{m_{\alpha,|\beta|} - m_{\alpha,0}} = \frac{m_{\alpha',0}}{m_{\alpha',\nu}}\frac{m_{\alpha',|\beta|} - m_{\alpha',\nu}}{m_{\alpha',|\beta|} - m_{\alpha',0}}.$$

PROOF. (1), (2), and (3) are immediately obtained from the second equality in Proposition 4.20. To prove $m_{\alpha,\nu} > 0$, note that the sign of d_α is equal to the sign of u_α. This implies that the sign of $d_{\mathfrak{p}(\alpha)}d_\alpha^{-1}$ is equal to the sign of everywhere non-zero function $\prod_{i \in \mathfrak{p}(\alpha)}(h_i + e_{\mathfrak{p}(\alpha)\alpha})$. Thus we have $m_{\alpha,\nu} > 0$. The remaining inequalities in (4) follows from the inequality
$$1 = c_{\mathfrak{p}(\alpha),0} > \cdots > c_{\mathfrak{p}(\alpha),|\mathfrak{p}(\alpha)|} = 0$$
and the fact that either $e_{\mathfrak{p}(\alpha)\alpha} > 0$ or $e_{\mathfrak{p}(\alpha)\alpha} < -1$. To prove (5), note that $d_\beta d_\alpha^{-1}$, $c_{\beta,\nu}$ $(1 \le \nu \le |\beta| - 1)$, and $e_{\beta\alpha}$ are uniquely determined from $m_{\alpha,\nu}$ $(0 \le \nu \le |\beta|)$, where $\alpha \in \mathfrak{n}(\beta)$. In particular we have
$$e_{\beta\alpha} = \frac{m_{\alpha,0}}{m_{\alpha,|\beta|} - m_{\alpha,0}}, \qquad c_{\beta,\nu} = \frac{m_{\alpha,0}}{m_{\alpha,\nu}}\frac{m_{\alpha,|\beta|} - m_{\alpha,\nu}}{m_{\alpha,|\beta|} - m_{\alpha,0}}. \tag{4.9}$$
Hence (5) follows. □

REMARK. If h_i $(i \in \beta)$ and $e_{\beta\alpha}$ $(\beta \prec \alpha)$ are replaced with $1 - h_{s(\beta)+t(\beta)-i}$ and $-1 - e_{\beta\alpha}$ respectively for a non-maximal β, then, (1) the order of $Y_{\beta,0}, \ldots, Y_{\beta,|\beta|}$ and $m_{\alpha,0}, \ldots, m_{\alpha,|\beta|}$ are reversed ($\alpha \in \mathfrak{n}(\beta)$), and (2) d_γ is replaced with $(-1)^{|\beta|}d_\gamma$ $(\beta \prec \gamma)$.

If $\mathcal{A}$ is totally ordered, it is therefore possible to choose $\{h_i\}$ and $\{e_{\beta\alpha}\}$ so that every $e_{\beta\alpha}$ is positive and $m_{\alpha,0} < \cdots < m_{\alpha,|\mathfrak{p}(\alpha)|}$ for every non-minimal α. But, in general it is impossible.

5. Properties as a toric variety

In the previous section we have proved that M is a toric variety with respect to the action of G. In this section we shall specify the fan of the toric variety M, and describe some properties that are useful for the "existence problem". Throughout this section we shall refer to Fulton [**Fu**] and Oda [**O1**] on the general theory for toric varieties.

The fan of M. As a toric variety, M is constructed from the lattice $\Gamma \subset \mathfrak{k}$ and a set Δ of polyhedral cones in the Lie algebra $\mathfrak{k}$. The pair (Γ, Δ) (or the set Δ if Γ is known) is called the fan of M. In our case, the invariant hypersurfaces $L_{\alpha,\nu}$ $((\alpha,\nu) \in \mathcal{J})$ and the information on their intersections will determine Δ. We first describe (Γ, Δ) in terms of the partially ordered set $\mathcal{A}$ and the numbers $|\alpha|$, $m_{\alpha,\nu}$, and then prove that it is the fan of M.

Let $\tilde{\mathfrak{k}}$ be the real vector space of dimension $n + \#\mathcal{A}$ equipped with the basis $\widetilde{Y}_{\alpha,\nu}$ $((\alpha,\nu) \in \mathcal{J})$, and let $\widetilde{\Gamma}$ be the lattice in $\tilde{\mathfrak{k}}$ generated by $\widetilde{Y}_{\alpha,\nu}$ $((\alpha,\nu) \in \mathcal{J})$. Define $\widetilde{Z}_\alpha \in \widetilde{\Gamma}$ $(\alpha \in \mathcal{A})$ by

$$\widetilde{Z}_\alpha = \sum_{\substack{\beta;\text{non-minimal} \\ \beta \preceq \alpha}} \left(\prod_{\beta \prec \gamma \preceq \alpha} m_{\gamma,0} \right) \sum_{\mu=1}^{|\mathfrak{p}(\beta)|} (m_{\beta,\mu} - m_{\beta,0}) \widetilde{Y}_{\mathfrak{p}(\beta),\mu}$$

if α is non-minimal, and by $\widetilde{Z}_\alpha = 0$ if α is minimal. Let Γ_0 be the subgroup of $\widetilde{\Gamma}$ generated by

$$R_\alpha = \sum_{\nu=0}^{|\alpha|} \widetilde{Y}_{\alpha,\nu} - \widetilde{Z}_\alpha \qquad (\alpha \in \mathcal{A}),$$

and let $\mathfrak{k}_0$ be the subspace of $\tilde{\mathfrak{k}}$ spanned by R_α $(\alpha \in \mathcal{A})$. Then we have the exact sequences

$$\begin{gathered} 0 \to \mathfrak{k}_0 \to \tilde{\mathfrak{k}} \xrightarrow{\rho} \mathfrak{k} \to 0, \\ 0 \to \Gamma_0 \to \widetilde{\Gamma} \xrightarrow{\rho} \Gamma \to 0, \end{gathered} \tag{5.1}$$

where ρ is the homomorphism defined by $\rho(\widetilde{Y}_{\alpha,\nu}) = Y_{\alpha,\nu}$ $((\alpha,\nu) \in \mathcal{J})$. Namely, Γ and $\mathfrak{k}$ are isomorphic to $\widetilde{\Gamma}/\Gamma_0$ and $\tilde{\mathfrak{k}}/\mathfrak{k}_0$ respectively.

For each $\iota \in \mathcal{I}$, let σ_ι be the n-dimensional cone in $\mathfrak{k}$ generated by $Y_{\alpha,\nu}$ $((\alpha,\nu) \in \mathcal{J}(\iota))$, i.e.,

$$\sigma_\iota = \{ \sum_{(\alpha,\nu)\in\mathcal{J}(\iota)} a_{\alpha,\nu} Y_{\alpha,\nu} \mid a_{\alpha,\nu} \geq 0 \}.$$

Let Δ be the set of the cones σ_ι $(\iota \in \mathcal{I})$ and all the faces of them. Here, a face of the cone σ_ι means a cone σ generated by $Y_{\alpha,\nu}$ $((\alpha,\nu) \in \mathcal{J}_0)$ for some subset $\mathcal{J}_0 \subset \mathcal{J}(\iota)$. Hence the assignment $\sigma \to \mathcal{J}_0$ gives the one-to-one correspondence between the cones in Δ and the subsets of $\mathcal{J}$ contained in some $\mathcal{J}(\iota)$. The 0-cone $\{0\}$ is supposed to be the face of every cone, which corresponds to the empty subset of $\mathcal{J}$. It is easily seen that (Γ, Δ) satisfies the conditions that a fan should satisfy, and the resulting toric variety, denoted by $X(\Delta)$, is compact and non-singular (see [**Fu**] Sections 2.4 and 2.5).

Notice that the construction above only needs a partially ordered set $\mathcal{A}$, integers $|\alpha|$ $(\alpha \in \mathcal{A})$, $m_{\alpha,\nu} - m_{\alpha,0}$ $(\alpha;\text{non-minimal},\ 0 \leq \nu \leq |\mathfrak{p}(\alpha)|)$, and rational numbers $m_{\alpha,0}$ $(\alpha,\ \mathfrak{p}(\alpha);\text{non-minimal})$ satisfying Proposition 1.8 (2) and Proposition 4.21. Namely, only the differences $m_{\alpha,\nu} - m_{\alpha,0}$ are used for such α that $\mathfrak{p}(\alpha)$ is minimal, and the condition (3) of Proposition 1.8 on the integers $|\alpha|$ for maximal α are not necessary for the construction. Generally, fans and toric varieties obtained in such a way from those data will be called *of KL type*. If Proposition 1,8 (3) is satisfied, then they will be called *of KL-A type*. If not, then they will be called *of KL-B type*.

REMARK. Since the fan of a toric variety is unique (cf. **[O2]** Theorem 4.1), and since Δ determines the sets of elements $\{Y_{\alpha,\nu}\}$ and $\{Z_\alpha\}$ of Γ, it follows that the partially ordered set $\mathcal{A}$ and the integers $|\alpha|$ ($\alpha \in \mathcal{A}$) are uniquely associated with a toric variety of KL type. Also, for each α there are only two possibilities for the ordering of $Y_{\alpha,0}, \ldots, Y_{\alpha,|\alpha|}$ so that the corresponding numbers $m_{\alpha,\nu}$ satisfy Proposition 4.21 (4); the alternative is the reversed order.

PROPOSITION 5.1. *(Γ, Δ) is the fan of M.*

Proposition 5.1 will be proved by giving an explicit identification of the toric variety $X(\Delta)$ (of KL-A type) with M. So, let us first review the construction of $X(\Delta)$.

For each $\sigma \in \Delta$ we define the semigroup $\mathcal{S}_\sigma$ by

$$\mathcal{S}_\sigma = \{\eta \in \Gamma^* \mid <\eta, Y> \geq 0 \text{ for any } Y \in \sigma \cap \Gamma\},$$

where Γ^* denotes the dual lattice of Γ. Regarding $\boldsymbol{C}$ as the multiplicative semigroup, we put

$$U_\sigma = \{u : \mathcal{S}_\sigma \to \boldsymbol{C}, \text{ a semigroup homomorphism}\}.$$

Here, homomorphisms are assumed to map unit to unit. If $\tau \subset \sigma$, then it is easily seen that

$$\mathcal{S}_\sigma \subset \mathcal{S}_\tau, \quad U_\tau \subset U_\sigma.$$

For the 0-cone, $\mathcal{S}_{\{0\}} = \Gamma^*$ and

$$U_{\{0\}} = \{u : \Gamma^* \to \boldsymbol{C}^\times, \text{ a group homomorphism}\} = \Gamma \otimes_{\boldsymbol{Z}} \boldsymbol{C}^\times,$$

which is an algebraic torus isomorphic to $(\boldsymbol{C}^\times)^n$. We shall denote it by $\mathcal{T}_\Gamma$. The group $\mathcal{T}_\Gamma$ naturally acts on each U_σ by

$$(u_0 u)(\eta) = u_0(\eta) u(\eta), \qquad u_0 \in \mathcal{T}_\Gamma,\ u \in U_\sigma,\ \eta \in \mathcal{S}_\sigma.$$

It follows from the definition that U_σ is an affine variety with coordinate ring $\boldsymbol{C}[\mathcal{S}_\sigma]$ (the semigroup ring). If the cone σ is k-dimensional, then U_σ is isomorphic to $\boldsymbol{C}^k \times (\boldsymbol{C}^\times)^{n-k}$. Then $X(\Delta)$ is obtained by gluing all U_σ with the relations

$$U_\sigma \supset U_{\sigma\cap\tau} \subset U_\tau.$$

The action of $\mathcal{T}_\Gamma$ on $X(\Delta)$ is also well-defined.

PROOF OF PROPOSITION 5.1. Fix a point $p_0 \in M^1$. Let $\iota \subset \mathcal{I}$, and let $(z^{(\iota)}_{\alpha,\nu};\ (\alpha,\nu) \in \mathcal{J}(\iota))$ be the coordinate system on $M - \cup_{\alpha\in\mathcal{A}} L_{\alpha,\iota(\alpha)}$ given in Corollary 4.16. Also, let $\eta^{(\iota)}_{\alpha,\nu}$ ($(\alpha,\nu) \in \mathcal{J}(\iota)$) be the basis of Γ^* dual to $Y_{\alpha,\nu}$ ($(\alpha,\nu) \in \mathcal{J}(\iota)$) (note that they are generators for $\mathcal{S}_{\sigma_\iota}$). Then, there is a natural holomorphic isomorphism

$$U_{\sigma_\iota} \to M - \cup_{\alpha\in\mathcal{A}} L_{\alpha,\iota(\alpha)} \quad (u \mapsto p) \tag{5.2}$$

given by

$$u(\eta^{(\iota)}_{\alpha,\nu}) = z^{(\iota)}_{\alpha,\nu}(p) \quad ((\alpha,\nu) \in \mathcal{J}(\iota)). \tag{5.3}$$

If $\sigma \in \Delta$ is a face of σ_ι, then the mapping (5.2) gives the holomorphic isomorphism

$$U_\sigma \to M - \cup_{(\alpha,\nu)\in\mathcal{J}-\mathcal{J}_0} L_{\alpha,\nu}, \tag{5.4}$$

where $\mathcal{J}_0$ is the subset of $\mathcal{J}$ corresponding to σ;

$$\mathcal{J}_0 = \{(\alpha, \nu) \in \mathcal{J} \mid Y_{\alpha,\nu} \in \sigma\} \subset \mathcal{J}(\iota).$$

It is easily seen that the isomorphism (5.4) is independent of the choice of σ_ι containing σ. Hence we obtain the holomorphic isomorphism $X(\Delta) \to M$.

Defining the isomorphism $\mathcal{T}_\Gamma \to G$ by

$$Y_{\alpha,\nu} \otimes e^{t+\sqrt{-1}s} \mapsto \exp(-tIY_{\alpha,\nu} + sY_{\alpha,\nu}),$$

we can easily see that the isomorphism $X(\Delta) \to M$ commutes with the actions of the groups $\mathcal{T}_\Gamma$ and G. This completes the proof. □

From now on, we shall fix a point $p_0 \in M^1$ (the base point) and identify each U_σ with the subset $M - \cup_{(\alpha,\nu)\in\mathcal{J}-\mathcal{J}_0} L_{\alpha,\nu}$ of M by the isomorphism given in the proof of Proposition 5.1, where $\mathcal{J}_0 \subset \mathcal{J}$ corresponds to σ. Also, the group $\mathcal{T}_\Gamma$ will be identified with G. For $\sigma \in \Delta$ corresponding to $\mathcal{J}_0$ we put

$$O_\sigma = U_\sigma \cap \cap_{(\alpha,\nu)\in\mathcal{J}_0} L_{\alpha,\nu}.$$

Then, O_σ is a G-orbit isomorphic to $(\boldsymbol{C}^\times)^{n-k}$ ($k = \dim\sigma$), and

$$M = \cup_{\sigma\in\Delta} O_\sigma, \quad U_\sigma = \cup_{\tau\subset\sigma} O_\tau.$$

Note that $O_{\{0\}} = M^1$. Let $V(\sigma)$ be the closure of O_σ in M. Then we have

$$V(\sigma) = \cap_{(\alpha,\nu)\in\mathcal{J}_0} L_{\alpha,\nu} = \cup_{\tau\supset\sigma} O_\tau$$

(see **[Fu]** Section 3.1).

Since Δ contains n-dimensional cones, we have:

COROLLARY 5.2. *M is simply connected.*

For the proof, see **[Fu]** p. 56, Proposition. Actually, one can see more about the topology of M: There is a cell-decomposition of M consisting of $\prod_\alpha(|\alpha| + 1)$ even-dimensional cells; the number of $2k$-dimensional cells is equal to the number of $\iota \in \mathcal{I}$ such that $\sum_\alpha \iota(\alpha) = k$. This is made in a similar way as **[Fu]** pp. 101-103. Since the result is not used in this paper, we omit the detail.

Fibre bundles associated with M. In general, let (Γ_i, Δ_i) ($i = 1, 2$) be two fans, and let $\phi : \Gamma_1 \to \Gamma_2$ be a homomorphism such that the induced linear homomorphism

$$\phi : \Gamma_1 \otimes_{\boldsymbol{Z}} \boldsymbol{R} \to \Gamma_2 \otimes_{\boldsymbol{Z}} \boldsymbol{R}$$

maps each $\sigma \in \Delta_1$ into some $\sigma' \in \Delta_2$. Denoting the resulting toric varieties by $X(\Delta_i)$, we have

PROPOSITION 5.3. *There is a natural holomorphic mapping $\phi_\sharp : X(\Delta_1) \to X(\Delta_2)$ that is equivariant with respect to the naturally induced homomorphism*

$$\phi_\sharp : \mathcal{T}_{\Gamma_1} \to \mathcal{T}_{\Gamma_2}$$

of algebraic tori.

For the proof, see **[O1]** p. 19, Theorem 1.13 (see also **[Fu]** p. 41, Exercise). As applications of this general result, we shall obtain two kinds of fibre bundles associated with M. We now explain the first one. Let $\widetilde{\Gamma}$ and $\widetilde{\mathfrak{k}}$ be as in the previous subsection. For each $\sigma \in \Delta$ corresponding to $J_0 \subset \mathcal{J}$, define the cone $\widetilde{\sigma}$ in $\widetilde{\mathfrak{k}}$ by

$$\widetilde{\sigma} = \{ \sum_{(\alpha,\nu)\in\mathcal{J}_0} a_{\alpha,\nu}\widetilde{Y}_{\alpha,\nu} \mid a_{\alpha,\nu} \geq 0\},$$

and put $\widetilde{\Delta} = \{\widetilde{\sigma} \mid \sigma \in \Delta\}$. As is easily verified, $(\widetilde{\Gamma}, \widetilde{\Delta})$ is a fan. Then, by Proposition 5.3 the homomorphism ρ induces the equivariant holomorphic mapping $\rho_\sharp : X(\widetilde{\Delta}) \to M$.

PROPOSITION 5.4. (1) *The toric variety $X(\widetilde{\Delta})$ and the algebraic torus $\mathcal{T}_{\widetilde{\Gamma}}$ are naturally isomorphic to*

$$\prod_{\alpha\in\mathcal{A}} (\boldsymbol{C}^{|\alpha|+1} - \{0\}) = \{(z_\alpha;\ \alpha \in \mathcal{A}) \mid z_\alpha = (z_{\alpha,0}, \ldots, z_{\alpha,|\alpha|}) \in \boldsymbol{C}^{|\alpha|+1} - \{0\}\}$$

and

$$(\boldsymbol{C}^\times)^{n+\#\mathcal{A}} = \{(\lambda_{\alpha,\nu};\ (\alpha,\nu) \in \mathcal{J}) \mid \lambda_{\alpha,\nu} \in \boldsymbol{C}^\times\}$$

respectively.

(2) *$\rho_\sharp : X(\widetilde{\Delta}) \to M$ is a principal fibre bundle with structure group $\mathcal{T}_{\Gamma_0}$.*

PROOF. (1) Let $\widetilde{Y}^*_{\alpha,\nu}$ $((\alpha,\nu) \in \mathcal{J})$ be the basis of $\widetilde{\Gamma}^*$ dual to $\widetilde{Y}_{\alpha,\nu}$ $((\alpha,\nu) \in \mathcal{J})$. Then, all the semigroups $\mathcal{S}_{\widetilde{\sigma}}$ contain the semigroup generated by $\widetilde{Y}^*_{\alpha,\nu}$ $((\alpha,\nu) \in \mathcal{J})$. This implies that all the affine varieties $U_{\widetilde{\sigma}}$ are realized in $\boldsymbol{C}^{n+\#\mathcal{A}}$;

$$\boldsymbol{C}^{n+\#\mathcal{A}} = \{(z_{\alpha,\nu};\ (\alpha,\nu) \in \mathcal{J})\},$$
$$U_{\widetilde{\sigma}} = \{(z_{\alpha,\nu}) \mid z_{\alpha,\nu} \neq 0 \text{ for } (\alpha,\nu) \notin \mathcal{J}_0\},$$

$\mathcal{J}_0$ corresponding to σ. From this the assertion easily follows.

(2) We first review how the mapping $\rho_\sharp$ is constructed: The surjective homomorphism $\rho : \widetilde{\Gamma} \to \Gamma$ induces the inclusion $\rho^* : \Gamma^* \to \widetilde{\Gamma}^*$. This gives the inclusion $\mathcal{S}_\sigma \to \mathcal{S}_{\widetilde{\sigma}}$ for any $\sigma \in \Delta$. Thus the mapping $\rho_\sharp : U_{\widetilde{\sigma}} \to U_\sigma$ is defined by

$$u \mapsto u|_{\mathcal{S}_\sigma} \qquad (u \in U_{\widetilde{\sigma}}).$$

Now, fix $\iota \in \mathcal{I}$ and define a splitting $j_\iota : \Gamma \to \widetilde{\Gamma}$ of the exact sequence (5.1) by $j_\iota(Y_{\alpha,\nu}) = \widetilde{Y}_{\alpha,\nu}$ $((\alpha,\nu) \in \mathcal{J}(\iota))$. Let Γ_ι be its image. Then we have the direct sum decompositions

$$\widetilde{\Gamma} = \Gamma_0 + \Gamma_\iota, \qquad \widetilde{\Gamma}^* = \Gamma_\iota^\perp + \Gamma^*,$$

and accordingly,

$$\mathcal{S}_{\widetilde{\sigma}_\iota} = \Gamma_\iota^\perp + \mathcal{S}_{\sigma_\iota}.$$

Since $\Gamma_\iota^\perp$ is identified with Γ_0^* by the natural homomorphism $\widetilde{\Gamma}^* \to \Gamma_0^*$, we thus obtain the holomorphic isomorphism

$$U_{\widetilde{\sigma}_\iota} \to \mathcal{T}_{\Gamma_0} \times U_{\sigma_\iota} \qquad (u \mapsto (u|_{\Gamma_\iota^\perp}, u|_{\mathcal{S}_{\sigma_\iota}})). \tag{5.5}$$

Clearly, the mapping above also gives the isomorphism $\mathcal{T}_{\widetilde{\Gamma}} \to \mathcal{T}_{\Gamma_0} \times G$ of algebraic tori, with which the isomorphism (5.5) is equivariant. This proves (2). □

Now, let us explain the other kind of fibre bundles that are naturally associated with M. For convenience we shall introduce a topology on the set $\mathcal{A}$: A subset $\mathcal{B}$ of $\mathcal{A}$ is open if it possesses the property;

$$\text{if } \beta \in \mathcal{B} \text{ and } \gamma \preceq \beta, \text{ then } \gamma \in \mathcal{B}. \tag{5.6}$$

Let $\mathcal{A}'$ be an open subset of $\mathcal{A}$, and put $\mathcal{A}'' = \mathcal{A} - \mathcal{A}'$. Let Γ' be a subgroup of Γ generated by $Y_{\alpha,\nu}$ $((\alpha,\nu) \in \mathcal{J},\ \alpha \in \mathcal{A}')$, and let $\mathfrak{k}'$ be the subspace of $\mathfrak{k}$ spanned by those vectors. Put

$$\Gamma'' = \Gamma/\Gamma', \quad \mathfrak{k}'' = \mathfrak{k}/\mathfrak{k}',$$

and let $\pi : \mathfrak{k} \to \mathfrak{k}''$ be the natural projection. Also, put

$$\Delta' = \{\sigma \in \Delta \mid \sigma \subset \mathfrak{k}'\}, \quad \Delta'' = \{\pi(\sigma) \mid \sigma \in \Delta\}.$$

Then the pairs (Γ', Δ') and (Γ'', Δ'') become fans. It is clear that $X(\Delta'')$ is a toric variety of KL-A type. For $X(\Delta')$, it can be only said that it is of KL type. Since the homomorphism π satisfies the assumption of Proposition 5.3, we have the equivariant mapping

$$\pi_\sharp : M \to X(\Delta'').$$

PROPOSITION 5.5. *$\pi_\sharp : M \to X(\Delta'')$ is a fibre bundle with typical fibre $X(\Delta')$. More precisely, for each $\iota \in \mathcal{I}$ there is an isomorphism $G \to \mathcal{T}_{\Gamma'} \times \mathcal{T}_{\Gamma''}$ of algebraic tori and an equivariant holomorphic isomorphism*

$$\pi_\sharp^{-1}(U_{\pi(\sigma_\iota)}) \to X(\Delta') \times U_{\pi(\sigma_\iota)}.$$

PROOF. Fix $\iota \in \mathcal{I}$, and let Γ_1 be the subgroup of Γ generated by $Y_{\alpha,\nu}$ $((\alpha,\nu) \in \mathcal{J}(\iota),\ \alpha \in \mathcal{A}'')$. We then have the direct sum decompositions

$$\Gamma = \Gamma' + \Gamma_1, \quad \Gamma^* = \Gamma_1^\perp + (\Gamma'')^*,$$

and

$$\mathcal{S}_{\sigma_\iota} = \mathcal{S}_{\sigma'_\iota} + \mathcal{S}_{\pi(\sigma_\iota)},$$

where $\sigma'_\iota = \sigma_\iota \cap \mathfrak{k}' \in \Delta'$, and $\Gamma_1^\perp$ is identified with $(\Gamma')^*$ by the projection $\Gamma^* \to (\Gamma')^*$. This induces the holomorphic isomorphism

$$U_{\sigma_\iota} \to U_{\sigma'_\iota} \times U_{\pi(\sigma_\iota)}, \tag{5.7}$$

and the isomorphism

$$\mathcal{T}_\Gamma \to \mathcal{T}_{\Gamma'} \times \mathcal{T}_{\Gamma''}$$

of algebraic tori so that the mapping (5.7) is equivariant. Then, varying $\iota \in \mathcal{I}$ so that $\iota(\alpha)$ remains unchanged for any $\alpha \in \mathcal{A}''$, and taking the union of both sides of (5.7) with respect to all such ι, we have the equivariant isomorphism

$$\pi_\sharp : \pi^{-1}(U_{\pi(\sigma_\iota)}) \to X(\Delta') \times U_{\pi(\sigma_\iota)}.$$

□

Let $\mathcal{A} = \cup_{i=1}^k \mathcal{A}_i$ (disjoint union) be the decomposition of $\mathcal{A}$ into the connected components. Let Γ_i be the subgroup of Γ generated by $Y_{\alpha,\nu}$ $(\alpha \in \mathcal{A}_i)$, and let $\mathfrak{k}_i$ be the subspace of $\mathfrak{k}$ spanned by Γ_i. Clearly, Γ is the direct sum of those subgroups, and correspondingly,

$$G = \mathcal{T}_{\Gamma_1} \times \cdots \times \mathcal{T}_{\Gamma_k}.$$

Putting

$$\Delta_i = \{\sigma \cap \mathfrak{t}_i \mid \sigma \in \Delta\},$$

we obtain fans (Γ_i, Δ_i).

COROLLARY 5.6. *There is a natural holomorphic isomorphism*

$$M \to X(\Delta_1) \times \cdots \times X(\Delta_k)$$

that is equivariant with respect to the identification

$$G = \mathcal{T}_{\Gamma_1} \times \cdots \times \mathcal{T}_{\Gamma_k}.$$

Moreover, each $X(\Delta_i)$ naturally becomes a Kähler-Liouville manifold so that M becomes the product manifold as Kähler-Liouville manifold.

PROOF. The former half is an immediate consequence of Proposition 5.5. The latter half is then obvious. □

We now go back to the situation of Proposition 5.5 and observe the fibre bundle $\pi_\sharp : M \to X(\Delta'')$ from another point of view. Let $\widetilde{\Gamma}'$ and $\widetilde{\Gamma}''$ be the subgroups of $\widetilde{\Gamma}$ generated by $\widetilde{Y}_{\alpha,\nu}$ $((\alpha,\nu) \in \mathcal{J},\ \alpha \in \mathcal{A}')$ and $\widetilde{Y}_{\alpha,\nu}$ $((\alpha,\nu) \in \mathcal{J},\ \alpha \in \mathcal{A}'')$ respectively. We then have

$$\widetilde{\Gamma} = \widetilde{\Gamma}' + \widetilde{\Gamma}'' \quad \text{(direct sum)}. \tag{5.8}$$

Let $\widetilde{\pi} : \widetilde{\Gamma} \to \widetilde{\Gamma}''$ be the projection. The homomorphism $\rho : \widetilde{\Gamma} \to \Gamma$ induces the homomorphisms $\widetilde{\Gamma}' \to \Gamma'$ and $\rho'' : \widetilde{\Gamma}'' \to \Gamma''$ (the latter is given by $\pi \circ \rho$). Let Γ'_0 and Γ''_0 be the kernel of those homomorphisms. Thus we have the following commutative diagram whose rows and columns are exact:

$$\begin{array}{ccccccccc}
 & & 0 & & 0 & & 0 & & \\
 & & \downarrow & & \downarrow & & \downarrow & & \\
0 & \longrightarrow & \Gamma'_0 & \longrightarrow & \widetilde{\Gamma}' & \xrightarrow{\rho} & \Gamma' & \longrightarrow & 0 \\
 & & \downarrow & & \downarrow & & \downarrow & & \\
0 & \longrightarrow & \Gamma_0 & \longrightarrow & \widetilde{\Gamma} & \xrightarrow{\rho} & \Gamma & \longrightarrow & 0 \\
 & & \downarrow{\scriptstyle \widetilde{\pi}} & & \downarrow{\scriptstyle \widetilde{\pi}} & & \downarrow{\scriptstyle \pi} & & \\
0 & \longrightarrow & \Gamma''_0 & \longrightarrow & \widetilde{\Gamma}'' & \xrightarrow{\rho''} & \Gamma'' & \longrightarrow & 0 \\
 & & \downarrow & & \downarrow & & \downarrow & & \\
 & & 0 & & 0 & & 0 & &
\end{array}$$

Note that the splitting in the mid column does not induce the splitting in the left one, i.e., $\Gamma_0 \cap \widetilde{\Gamma}'' \neq \Gamma''_0$ in general. Let $\widetilde{\mathfrak{t}}''$ be the subspace of $\widetilde{\mathfrak{t}}$ spanned by $\widetilde{\Gamma}''$, and put

$$\widetilde{\Delta}'' = \{\sigma \in \widetilde{\Delta} \mid \sigma \subset \widetilde{\mathfrak{t}}''\}.$$

Then, $(\widetilde{\Gamma}'', \widetilde{\Delta}'')$ becomes a fan, and the homomorphism ρ'' induces the principal $\mathcal{T}_{\Gamma''_0}$-bundle $\rho''_\sharp : X(\widetilde{\Delta}'') \to X(\Delta'')$.

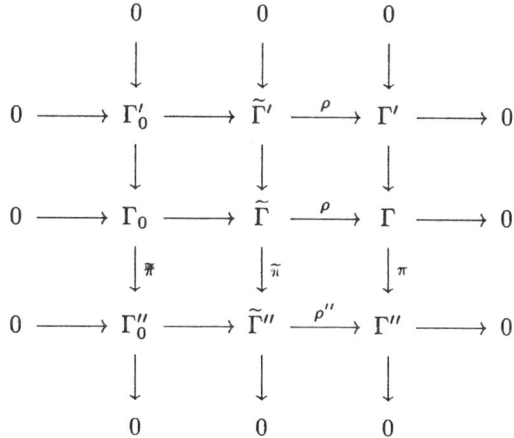

We now define the homomorphism $\psi : \Gamma_0'' \to \Gamma'$ as follows: Let $\widetilde{\Gamma} \to \widetilde{\Gamma}'$ be the projection with respect to the decomposition (5.8). Restricting it to Γ_0, we have the homomorphism $\Gamma_0 \to \widetilde{\Gamma}'$. The restriction of this mapping to Γ_0' being the identity, we thus obtain the homomorphism

$$\psi_1 : \Gamma_0'' = \Gamma_0/\Gamma_0' \to \widetilde{\Gamma}'/\Gamma_0' = \Gamma'.$$

We put $\psi = -\psi_1$. The following formula is easily obtained:

$$\psi(\widetilde{\pi}(R_\alpha)) = \left(\prod_{\alpha_0 \prec \beta \preceq \alpha} m_{\beta,0} \right) Z_{\alpha_0} \qquad (\alpha \in \mathcal{A}''),$$

where α_0 is the minimal element of $\mathcal{A}''$ satisfying $\alpha_0 \preceq \alpha$. The induced homomorphism $\mathcal{T}_{\Gamma_0''} \to \mathcal{T}_{\Gamma'}$ of algebraic tori is also denoted by ψ. Through this homomorphism $\mathcal{T}_{\Gamma_0''}$ acts on $X(\Delta')$.

PROPOSITION 5.7. *The fibre bundle $\pi_\sharp : M \to X(\Delta'')$ is isomorphic to the fibre product*

$$X(\widetilde{\Delta}'') \times_{\mathcal{T}_{\Gamma_0''}} X(\Delta') \to X(\Delta'').$$

PROOF. Let $\widetilde{\mathfrak{k}}'$ be the subspace of $\widetilde{\mathfrak{k}}$ spanned by $\widetilde{\Gamma}'$, and put $\widetilde{\Delta}' = \{\sigma \in \widetilde{\Delta} \mid \sigma \subset \widetilde{\mathfrak{k}}'\}$. Then $(\widetilde{\Gamma}', \widetilde{\Delta}')$ is also a fan, and we have

$$X(\widetilde{\Delta}) = X(\widetilde{\Delta}'') \times X(\widetilde{\Delta}').$$

Hence $M = X(\widetilde{\Delta})/\mathcal{T}_{\Gamma_0}$ is equal to

$$((X(\widetilde{\Delta}'') \times X(\widetilde{\Delta}'))/\mathcal{T}_{\Gamma_0'})/\mathcal{T}_{\Gamma_0''} = (X(\widetilde{\Delta}'') \times X(\Delta'))/\mathcal{T}_{\Gamma_0''}.$$

Since the action of $\mathcal{T}_{\Gamma_0''}$ on $X(\widetilde{\Delta}'') \times X(\Delta')$ is given by

$$(p, q)g = (pg, \psi_1(g)q) = (pg, \psi(g^{-1})q),$$

the proposition follows. □

Line bundles. Let $\mathrm{Pic}(M)$ denote the group of the isomorphism classes of holomorphic line bundles over M. To each $\xi \in \widetilde{\Gamma}^*$ we associate a divisor of M;

$$\xi \mapsto \sum_{(\alpha,\nu)\in\mathcal{J}} < \xi, \widetilde{Y}_{\alpha,\nu} > L_{\alpha,\nu}.$$

where $<,>$ denotes the natural pairing of $\widetilde{\Gamma}^*$ and $\widetilde{\Gamma}$. Let Q_ξ be the line bundle over M associated with this divisor. Then we have the homomorphism $\widetilde{\Gamma}^* \to \mathrm{Pic}(M)$ $(\xi \mapsto Q_\xi)$. Also, the homomorphism $-\xi : \widetilde{\Gamma} \to \boldsymbol{Z}$ induces the homomorphism $\chi_{-\xi} : \mathcal{T}_{\widetilde{\Gamma}} \to \boldsymbol{C}^\times$. Restricting it to $\mathcal{T}_{\Gamma_0}$, we have another line bundle over M associated with $\pi_\sharp : X(\widetilde{\Delta}) \to M$.

PROPOSITION 5.8. (1) *The following sequence is exact:*

$$0 \to \Gamma^* \xrightarrow{\rho^*} \widetilde{\Gamma}^* \to \mathrm{Pic}(M) \to 0.$$

(2) Q_ξ *is isomorphic to the fibre product* $X(\widetilde{\Delta}) \times_{\chi_{-\xi}} \boldsymbol{C}$, *where* $\chi_{-\xi}$ *is regarded as the homomorphism* $\mathcal{T}_{\Gamma_0} \to \boldsymbol{C}^\times$ *by restriction.*

(3) *The assignment of the first Chern class* $c_1(Q)$ *to each* $Q \in \mathrm{Pic}(M)$ *gives the isomorphism* $\mathrm{Pic}(M) \to H^2(M, \boldsymbol{Z})$.

PROOF. For (1) and (3), see [**Fu**] pp. 63-64 and [**O1**] Corollary 2.5. (2) is easy. □

Put

$$\zeta_\alpha = c_1(Q_{\widetilde{Y}^*_{\alpha,0}}) \in H^2(M, \boldsymbol{Z}) \qquad (\alpha \in \mathcal{A}).$$

The proposition above implies that the elements ζ_α $(\alpha \in \mathcal{A})$ form a basis of $H^2(M, \boldsymbol{Z})$. Its dual basis is given as follows: Let $\tau(\alpha) \in \Delta$ be the $(n-1)$-dimensional cone generated by

$$\{Y_{\beta,\nu} \mid 1 \le \nu \le |\beta| \text{ if } \beta \ne \alpha;\ 2 \le \nu \le |\beta| \text{ if } \beta = \alpha\}.$$

Then $V(\tau(\alpha))$ is 1-dimensional (isomorphic to $\boldsymbol{C}P^1$). Let $[V(\tau(\alpha))]$ denote its fundamental class in $H_2(M, \boldsymbol{Z})$.

PROPOSITION 5.9. $< \zeta_\alpha,\ [V(\tau(\beta))] >= \delta_{\alpha\beta}$, *where* $<,>$ *denotes the natural pairing of* $H^2(M, \boldsymbol{Z})$ *and* $H_2(M, \boldsymbol{Z})$.

PROOF. The assertion easily follows from the fact:

$$L_{\alpha,0} \cap V(\tau(\beta)) = \begin{cases} 0 & \text{if } \alpha \ne \beta \\ V(\sigma_\iota) = \{\text{a point}\} & \text{if } \alpha = \beta, \end{cases}$$

where $\iota \in \mathcal{I}$ is given by $\iota(\gamma) = \delta_{\gamma\alpha}$. □

The next theorem specifies the cohomology class $[\omega] \in H^2(M, \boldsymbol{R})$ of the Kähler form ω. Let $\mathcal{A}_i$ $(1 \le i \le k)$ be the connected components of $\mathcal{A}$, and let α_i denote the unique minimal element of $\mathcal{A}_i$.

THEOREM 5.10.

(1)

$$\int_{V(\tau(\alpha))} \omega = \frac{2\pi}{d_\alpha} \prod_{\beta \prec \alpha} \prod_{\nu=1}^{|\beta|} (c_{\beta,\nu} + e_{\beta\alpha}).$$

(2)

$$[\omega] = \sum_{i=1}^{k} \frac{2\pi}{d_{\alpha_i}} \left(\sum_{\alpha \succ \alpha_i} \left(\prod_{\alpha_i \prec \beta \preceq \alpha} m_{\beta,0} \right) \zeta_\alpha + \zeta_{\alpha_i} \right).$$

PROOF. (1) As is easily seen, the action of the circle group $\{\exp(tY_{\alpha,0})\}$ ($t \in \boldsymbol{R}/2\pi\boldsymbol{Z}$) on $V(\tau(\alpha))$ has the two fixed points q_0 and q_1;

$$\{q_0\} = V(\tau(\alpha)) \cap V(\boldsymbol{R}_{\geq 0}Y_{\alpha,0}), \qquad \{q_1\} = V(\tau(\alpha)) \cap V(\boldsymbol{R}_{\geq 0}Y_{\alpha,1}),$$

where $\boldsymbol{R}_{\geq 0}$ denotes the set of non-negative real numbers. Let $\gamma(s)$ $(0 \leq s \leq l)$ be a geodesic of unit speed on $V(\tau(\alpha))$ from q_0 to q_1. Then, parametrizing $V(\tau(\alpha))$ by (s,t), we have

$$\int_{V(\tau(\alpha))} \omega = \int_0^l \int_0^{2\pi} \omega(\frac{\partial}{\partial s}, \frac{\partial}{\partial t}) dt ds,$$

and

$$\omega(\frac{\partial}{\partial s}, \frac{\partial}{\partial t}) = d_{\alpha,0}^{-1} \frac{d}{ds} v_\alpha(c_{\alpha,0})(\gamma(s)).$$

Hence

$$\int_{V(\tau(\alpha))} \omega = \frac{2\pi}{d_{\alpha,0}} \left(v_\alpha(c_{\alpha,0})(q_1) - v_\alpha(c_{\alpha,0})(q_0)\right).$$

Since $v_\alpha(c_{\alpha,0})(q_0) = 0$ and

$$v_\alpha(c_{\alpha,0})(q_1) = \left(\prod_{\beta \prec \alpha} \prod_{\nu=1}^{|\beta|} (c_{\beta,\nu} + e_{\beta\alpha}) \right) \prod_{\mu=1}^{|\alpha|} (c_{\alpha,\mu} - c_{\alpha,0}),$$

the assertion follows. (2) is immediately obtained from (1). □

6. Bundle structure associated with a subset of $\mathcal{A}$

In the previous section we have proved that an open subset $\mathcal{A}'$ induces the fibre bundle $\pi_\sharp : M \to X(\Delta'')$ whose typical fibre is $X(\Delta')$. In this section we shall show that the toric varieties $X(\Delta')$ and $X(\Delta'')$ naturally possess structures of Kähler-Liouville manifold inherited from M. Since the numbering $i = 1, \dots, n$ is inconvenient for the purpose of this section, we shall use (α, ν) $(\alpha \in \mathcal{A}, 1 \leq \nu \leq |\alpha|)$ instead. The correspondence is given by

$$(\alpha, \nu) \leftrightarrow i = s(\alpha) + \nu - 1.$$

Fix an open subset $\mathcal{A}'$, and let (Γ', Δ') and (Γ'', Δ'') be as in the previous section. TM is naturally decomposed to the sum of (mutually orthogonal) two subbundles; $TM = D' + D''$, where

$$D' = \sum_{\alpha \in \mathcal{A}'} D_\alpha, \quad D'' = \sum_{\alpha \in \mathcal{A}''} D_\alpha.$$

Clearly, D' is integrable, and the maximal integral submanifolds are the fibres of $\pi_\sharp : M \to X(\Delta'')$. Let $\mathcal{A}_1'', \dots, \mathcal{A}_r''$ be the connected components of $\mathcal{A}''$, and let α_s be the (unique) minimal element of A_s'' $(1 \leq s \leq r)$. Put

$$D_s'' = \sum_{\alpha \in \mathcal{A}_s''} D_\alpha \qquad (1 \leq s \leq r).$$

Recalling the orthonormal frame $V_{\alpha,\nu}, IV_{\alpha,\nu}$ $(\alpha \in \mathcal{A}, 1 \leq \nu \leq |\alpha|)$ over M^1, we put

$$V_{\alpha,\nu}'' = \sqrt{|u_{\alpha_s}|} V_{\alpha,\nu} \qquad (\alpha \in \mathcal{A}_s'').$$

The following lemma is easily obtained by using the properties of the vector fields $W_{\alpha,\nu}$ (cf. Proposition 1.4).

LEMMA 6.1. *For $\alpha \in \mathcal{A}''$ and $\beta \in \mathcal{A}'$,*

$$[V''_{\alpha,\nu}, V_{\beta,\mu}] = [V''_{\alpha,\nu}, IV_{\beta,\mu}] = [IV''_{\alpha,\nu}, V_{\beta,\mu}] = [IV''_{\alpha,\nu}, IV_{\beta,\mu}] = 0.$$

Let us recall the polynomial $F_\alpha(\lambda)$ in the indeterminate λ whose coefficients are elements of $\mathcal{F}$ (cf. Section 2):

(6.1)

$$F_\alpha(\lambda) = |u_\alpha| \sum_{\nu=1}^{|\alpha|} \prod_{\substack{1\le\mu\le|\alpha| \\ \mu\ne\nu}} (h_{\alpha,\mu} - \lambda) \cdot (V^2_{\alpha,\nu} + (IV_{\alpha,\nu})^2)$$

$$+ |u_\alpha| \sum_{\beta\in\mathfrak{n}(\alpha)} \frac{\prod_{\nu=1}^{|\alpha|}(h_{\alpha,\nu} + e_{\alpha\beta}) - \prod_{\nu=1}^{|\alpha|}(h_{\alpha,\nu} - \lambda)}{e_{\alpha\beta} + \lambda} \sum_{\gamma\succeq\beta} \sum_{\mu=1}^{|\gamma|} (V^2_{\gamma,\mu} + (IV_{\gamma,\mu})^2).$$

This polynomial is uniquely determined if the fundamental functions and the conjunction constants are specified. We shall call it the *generating polynomial.* The next proposition is an immediate consequence of the lemma above.

PROPOSITION 6.2. (1) *The vector fields $V''_{\alpha,\nu}$ are $\mathcal{T}_{\Gamma'}$-invariant.*
(2) *The horizontal subbundle D'' is $\mathcal{T}_{\Gamma'}$-invariant.*
(3) *For any $\alpha \in \mathcal{A}''$, the coefficients of $F_\alpha(\lambda)$ are $\mathcal{T}_{\Gamma'}$-invariant, and are sections of S^2D''.*

By virtue of Proposition 6.2 (3), the coefficients of $(\pi_\sharp)_* F_\alpha(\lambda)$ $(\alpha \in \mathcal{A}'')$ are well-defined sections of $S^2TX(\Delta'')$. Let $\mathcal{F}''$ be the vector space spanned by those sections. Also, the riemannian metric g'' on $X(\Delta'')$ is defined by the conditions: The subbundles D''_s $(1 \le s \le r)$ are mutually orthogonal with respect to $\pi_\sharp^* g''$, and

$$\pi_\sharp^* g'' = |u_{\alpha_s}|^{-1} g \quad \text{on } D''_s \qquad (1 \le s \le r).$$

It is easily seen that g'' is a Kähler metric. We denote by M'' the Kähler manifold $(X(\Delta''), g'')$.

THEOREM 6.3. *$(M'', \mathcal{F}'')$ is a Kähler-Liouville manifold of type* (A). *It possesses the following properties:*
(1) *The associated partially ordered set is naturally identified with $\mathcal{A}''$;*
(2) *the underlying toric variety is identical with $X(\Delta'')$;*
(3) *the fundamental functions $\{h''_{\alpha,\nu}\}$ $(\alpha \in \mathcal{A}'',\ 1 \le \nu \le |\alpha|)$ are given by $\pi_\sharp^*(h''_{\alpha,\nu}) = h_{\alpha,\nu}$;*
(4) *the conjunction constants $e''_{\alpha\beta}$ $(\alpha \in \mathcal{A}'',\ \alpha \preceq \beta)$ are given by $e''_{\alpha\beta} = e_{\alpha\beta}$;*
(5) *the scaling constants d''_α $(\alpha \in \mathcal{A}'')$ are given by $d''_\alpha = \epsilon(\alpha_s) d_\alpha$ $(\alpha \in \mathcal{A}''_s)$, where $\epsilon(\alpha_s)$ is the sign of d_{α_s}.*

PROOF. The commutativity of $\mathcal{F}''$ with respect to the Poisson bracket follows from that of $\mathcal{F}$. Since maximal elements of $\mathcal{A}''$ are also maximal in $\mathcal{A}$, it follows that $|\alpha| \ge 2$ for any maximal element α of $\mathcal{A}''$. This implies that $(M'', \mathcal{F}'')$ is of type (A). The properties (1), ..., (5) are easily verified. □

Next, let us consider the fibre. Define the Kähler metric $g'(q)$ on the fibre $\pi_\sharp^{-1}(q)$ $q \in M''$ by restricting g. With this metric we regard $\pi_\sharp^{-1}(q)$ as a Kähler manifold. Also, we define $\mathcal{F}'(q)$ as follows: Each $F \in \mathcal{F}$ is a section of $S^2D' + S^2D''$; so, taking the S^2D'-component F' of F, we put

$$\mathcal{F}'(q) = \{F'|_{\pi_\sharp^{-1}(q)} \mid F \in \mathcal{F}\}.$$

THEOREM 6.4. (1) *$(\pi_\sharp^{-1}(q), \mathcal{F}'(q))$ is a Kähler-Liouville manifold for any $q \in M''$.*

(2) *Let $\widetilde{X}$ be the horizontal lift (i.e., the lift as a section of D'') of a vector field X on M''. Then the one-parameter group $\{\phi_t\}$ of transformations of M generated by $\widetilde{X}$ gives the automorphisms $\pi_\sharp^{-1}(q) \to \pi_\sharp^{-1}(\phi_t(q))$ of Kähler manifolds, and preserves F' for each $F \in \mathcal{F}$.*

PROOF. (1) Let $F'_\alpha(\lambda)$ $(\alpha \in \mathcal{A}')$ be the S^2D'-component of $F_\alpha(\lambda)$. We have

$$F_\alpha(\lambda) = F'_\alpha(\lambda) + \epsilon(\alpha) \sum_{\beta \in \mathfrak{n}(\alpha)} \frac{u_\beta - v_\alpha(\lambda)}{e_{\alpha\beta} + \lambda} \sum_{\substack{1 \le s \le r \\ \alpha_s \succeq \beta}} 2|u_{\alpha_s}|^{-1} \widetilde{E}''_s, \tag{6.2}$$

where $\widetilde{E}''_s$ is the $S^2D''_s$-components of the horizontal lift of the energy function E'' of M'', and $\epsilon(\alpha)$ denotes the sign of u_α. Then, taking the S^3D'-components of

$$0 = \{F_\alpha(\lambda), F_\alpha(\mu)\},$$

we obtain

$$0 = \{F'_\alpha(\lambda), F'_\alpha(\mu)\}.$$

Hence $\mathcal{F}'$ is commutative.

(2) is an immediate consequence of Lemma 6.1. □

The typical fibre $X(\Delta')$ is naturally identified with the fibre $\pi_\sharp^{-1}(\pi_\sharp(p_0))$ passing through the base point $p_0 \in M^1$. Denoting the Kähler manifold $\pi_\sharp^{-1}(\pi_\sharp(p_0))$ by M', and $\mathcal{F}'(\pi_\sharp(p_0))$ by $\mathcal{F}'$, we obtain a Kähler-Liouville manifold $(M', \mathcal{F}')$. Note that it is of type (A) if and only if every maximal element α of $\mathcal{A}'$ satisfies $|\alpha| \geq 2$. The following theorem is immediate.

THEOREM 6.5. *If $(M', \mathcal{F}')$ is of type (A), then it possesses the following properties:*

(1) *The associated partially ordered set is naturally identified with $\mathcal{A}'$;*
(2) *the underlying toric variety is isomorphic to $X(\Delta')$;*
(3) *the fundamental functions $\{h'_{\alpha,\nu}\}$ $(\alpha \in \mathcal{A}',\ 1 \le \nu \le |\alpha|)$ are given by the restriction of $h_{\alpha,\nu}$ to M';*
(4) *the conjunction constants $e'_{\alpha\beta}$ $(\beta \in \mathcal{A}',\ \alpha \preceq \beta)$ are given by $e'_{\alpha\beta} = e_{\alpha\beta}$;*
(5) *the scaling constants d'_α $(\alpha \in \mathcal{A}')$ are given by $d'_\alpha = d_\alpha$.*

In case $(M', \mathcal{F}')$ is not of type (A), then the structure of toric variety on M' may be external, i.e., not determined by $(M', \mathcal{F}')$ itself. Nevertheless, we have the following

PROPOSITION 6.6. (1) *The maximal compact subgroup K' of the algebraic torus $\mathcal{T}_{\Gamma'}$ acts on the Kähler manifold M' as automorphisms and preserves each element of $\mathcal{F}'$.*

(2) *The geodesic flow of M' is integrable by means of $\mathcal{F}'$ and the Lie algebra of K'.*

The proof is clear. We shall say that a compact Kähler-Liouville manifold is *of type* (B) if it can be realized as the fibre of a fibre bundle obtained from a compact Kähler-Liouville manifold of type (A) and an open subset of the associated partially ordered set, and if it is not of type (A). By the definition, it possesses a structure of toric variety of KL-B type (not necessarily unique). It is another type of Kähler-Liouville manifold whose geodesic flow is integrable. In this paper we shall not mention further about such a manifold except the 1-dimensional case (see Section 8).

In the rest of this section, we shall show that the Kähler-Liouville manifold $(M, \mathcal{F})$ can be reconstructed from the structure of toric variety on M and the Kähler-Liouville manifolds $(M'', \mathcal{F}'')$ and $(M', \mathcal{F}')$, provided $(M', \mathcal{F}')$ is of type (A). By virtue of Corollary 5.6, M'' is described as the product $M''_1 \times \cdots \times M''_r$ of Kähler-Liouville manifolds, corresponding to the decomposition of $\mathcal{A}''$ into the connected components. Let ω''_s denote the Kähler form of M''_s. Also, let α_s denote the unique minimal element of $\mathcal{A}''_s$.

Put

$$Q = \cup_{q \in M''} (\text{the unique open orbit of } \mathcal{T}_{\Gamma'} \text{ in the fibre } \pi_\sharp^{-1}(q)).$$

Then Q is open and dense in M, and $\pi_\sharp : Q \to M''$ is a principal $\mathcal{T}_{\Gamma'}$-bundle. Proposition 5.7 implies that this bundle is isomorphic to

$$X(\tilde{\Delta}'') \times_\psi \mathcal{T}_{\Gamma'} \to M''.$$

Since the horizontal subbundle D'' is $\mathcal{T}_{\Gamma'}$-invariant, it defines the connection on this principal bundle. Let θ be the connection form, and Θ the $\mathfrak{g}'$-valued 2-form on M'' so that $\pi_\sharp^* \Theta$ is the curvature form ($\mathfrak{g}'$ is the Lie algebra of $\mathcal{T}_{\Gamma'}$).

PROPOSITION 6.7.

$$\Theta = \sum_{s=1}^{r} d''_{\alpha_s} \omega''_s \otimes Z_{\alpha_s}.$$

PROOF. By Propositions 1.2 and 1.10 we have

$$[V_i, IV_i] \equiv -\text{sgrad}\,(\log |u_\alpha|) \qquad \text{mod } (D_i)$$

for $\alpha \in \mathcal{A}''$ and $i \in \alpha$. This implies

$$[V''_i, IV''_i] \equiv -\text{sgrad}\,|u_{\alpha_s}| \qquad \text{mod } (D'')$$

for any $\alpha \subset \mathcal{A}''_s$ and $i \in \alpha$. Since $|d_{\alpha_s}| = d''_{\alpha_s}$, we have

$$d\theta(V''_i, IV''_i) = d''_{\alpha_s} Z_{\alpha_s} \qquad (\alpha \in \mathcal{A}''_s,\ i \in \alpha).$$

Also, it is easily seen that

$$d\theta(V''_i, V''_j) = d\theta(V''_i, IV''_j) = d\theta(IV''_i, IV''_j) = 0$$

for any $\alpha, \beta \in \mathcal{A}''$ and $i \in \alpha$, $j \in \beta$ $(i \neq j)$. Hence the proposition follows. □

From now on, we forget the structure of Kähler-Liouville manifold, and only assume that $M = X(\Delta)$ is a toric variety of KL-A type. Let $\mathcal{A}$ be the associated partially ordered set, and let $\mathcal{A}'$ be an open subset of it. Put $\mathcal{A}'' = \mathcal{A} - \mathcal{A}'$. Then we have the fibre bundle $\pi_\sharp : M \to X(\Delta'')$ with typical fibre $X(\Delta')$ as before, and the principal $\mathcal{T}_{\Gamma'}$-bundle $\pi_\sharp : Q \to X(\Delta'')$ as above. Let $(M'', \mathcal{F}'')$ be a Kähler-Liouville manifold of type (A) whose underlying toric variety is isomorphic to $X(\Delta'')$. We shall identify M'' with $X(\Delta'')$. Let $\mathcal{A}_s$, α_s, and ω''_s $(1 \le s \le r)$ be as above. For each non-minimal $\alpha \in \mathcal{A}$, let l_α be the largest positive integer satisfying $l_\alpha^{-1} Z_\alpha \in \Gamma$.

LEMMA 6.8. (1) *For any* $\alpha \in \mathcal{A}''_s$,

$$\left(\prod_{\alpha_s \prec \beta \preceq \alpha} m_{\beta,0} \right) l_{\alpha_s} \in \boldsymbol{Z}.$$

(2)

$$\left[\frac{d''_{\alpha_s}}{2\pi} l_{\alpha_s} \omega''_s \right] \in H^2(M''_s, \boldsymbol{Z}).$$

PROOF. (1) By Proposition 4.20 we have

$$Z_\alpha \equiv \left(\prod_{\alpha_s \prec \beta \preceq \alpha} m_{\beta,0} \right) Z_{\alpha_s} \qquad \mathrm{mod}\ (\sum_{\beta \in \mathcal{A}''} \sum_{\nu=1}^{|\beta|} \boldsymbol{Z} Y_{\beta,\nu}).$$

Thus the assertion follows. (2) follows from (1) and Theorem 5.10 (2). □

PROPOSITION 6.9. *There is a unique connection form* θ *on the principal bundle* $\pi_\sharp : Q \to M''$ *such that*

(1) *the associated curvature form is given by* $\pi_\sharp^* \Theta$, *where*

$$\Theta = \sum_{s=1}^{r} d''_{\alpha_s} \omega''_s \otimes Z_{\alpha_s};$$

(2) *the horizontal distribution defined by the kernel of* θ *is invariant with respect to the complex structure* I.

Accordingly, the $\mathcal{T}_{\Gamma'}$*-invariant horizontal subbundle* D'' *of* TM *with respect to* $\pi_\sharp : M \to M''$ *is uniquely determined by* Θ *so that the connection* $D''|_Q$ *on* Q *satisfies the conditions above. The connection form* θ *and the subbundle* D'' *are invariant under the action of the maximal compact subgroup* K *of* $\mathcal{T}_\Gamma$.

PROOF. First, we shall prove the uniqueness. Let θ be a connection form satisfying the condition (1) and (2). θ is $\mathfrak{g}'$-valued, and here $\mathfrak{g}'$ is regarded as a real Lie algebra with the complex structure I. Now, we regard it as a complex Lie algebra by replacing I with $\sqrt{-1}$. We shall write $\widetilde{\theta}$ (resp. $\widetilde{\Theta}$) instead of θ (resp. Θ) when $\mathfrak{g}'$ are regarded as the complex Lie algebra. By extending it $\boldsymbol{C}$-linearly to $TQ \otimes \boldsymbol{C}$, $\widetilde{\theta}$ becomes a $(1,0)$-form. Since $\widetilde{\Theta}$ is a $(1,1)$-from, we have

$$\partial \widetilde{\theta} = 0, \quad \bar{\partial} \widetilde{\theta} = \pi_\sharp^* \widetilde{\Theta}.$$

This implies that if θ_1 is another connection form satisfying the conditions (1) and (2), then $\widetilde{\theta} - \widetilde{\theta}_1$ is a holomorphic 1-form, and is projectable. Hence there is a holomorphic 1-form μ on M'' such that $\widetilde{\theta} - \widetilde{\theta}_1 = \pi_\sharp^* \mu$. However, since M'' is a compact, simply connected Kähler manifold, we have $\mu = 0$. Thus it follows that $\theta = \theta_1$.

Next, we shall prove the existence. Let P_s be a hermitian line bundle over M''_s with the canonical hermitian connection form $\widetilde{\theta}_s$ whose first Chern form is equal to

$$-\frac{d''_{\alpha_s}}{2\pi} l_{\alpha_s} \omega''_s$$

(for the existence of such a hermitian line bundle, see [**Kob1**] p. 41, Proposition). Put $U_s = \{v \in P_s \mid |v| = 1\}$ and $U = \prod_{s=1}^r U_s$. Then U is a principal $U(1)^r$-bundle over $M'' = \prod_{s=1}^r M''_s$. Let $\phi : U(1)^r \to \mathcal{T}_{\Gamma'}$ be the homomorphism given by

$$(\lambda_1, \dots, \lambda_r) \mapsto \prod_{s=1}^r (l_{\alpha_s}^{-1} Z_{\alpha_s} \otimes \lambda_s).$$

Then we obtain the associated $\mathcal{T}_{\Gamma'}$-bundle $U \times_\phi \mathcal{T}_{\Gamma'} \to M''$.

LEMMA 6.10. *The principal bundle $U \times_\phi \mathcal{T}_{\Gamma'} \to M''$ is naturally identified with the bundle $\pi_\sharp : Q \to M''$.*

PROOF. Let $\chi_s : \Gamma''_0 \to \boldsymbol{Z}$ $(1 \le s \le r)$ be the homomorphism given by

$$\chi_s(\widetilde{\pi}(R_\alpha)) = \left(\prod_{\alpha_s \prec \beta \preceq \alpha} m_{\beta,0} \right) l_{\alpha_s}.$$

The associated homomorphism $\mathcal{T}_{\Gamma''_0} \to \boldsymbol{C}^\times$ is also denoted by χ_s. Then, by Proposition 5.8 and Theorem 5.10 we see that the line bundle P_s is isomorphic to the fibre product

$$X(\widetilde{\Delta}'') \times_{\chi_s} \boldsymbol{C} \to M''.$$

Moreover, denoting by K''_0 the maximal compact subgroup of $\mathcal{T}_{\Gamma''_0}$, we have

$$\psi|_{K''_0} = \phi \circ (\chi_1, \dots, \chi_r)|_{K''_0}.$$

Therefore the lemma follows. □

We now continue the proof of Proposition 6.9. The direct sum of the connection forms $\widetilde{\theta}_s$, restricted to U_s, is a connection form on U. Composing this with the Lie algebra homomorphism associated with ϕ, we obtain a connection form θ on the principal bundle $\pi_\sharp : Q \to M''$. Then we clearly have $d\theta = \pi_\sharp^* \Theta$.

Finally, we prove the K-invariance. Let $k \in K$. Then the pull-back $k^*\theta$ is a connection form with the same curvature, because Θ is preserved by the transformation of M'' induced from k. Hence by the uniqueness we have $k^*\theta = \theta$. This completes the proof. □

Now, we moreover assume that there is a Kähler-Liouville manifold $(M', \mathcal{F}')$ of type (A) whose underlying toric variety is $X(\Delta')$. Then we have the following

THEOREM 6.11. *There is a unique Kähler-Liouville manifold $(M, \mathcal{F})$ of type* (A) *satisfying the following conditions:*

(1) *The underlying structure of toric variety is identical with the given one;*
(2) *the given Kähler-Liouville manifolds $(M', \mathcal{F}')$ and $(M'', \mathcal{F}'')$ are isomorphic with the ones induced from $(M, \mathcal{F})$.*

PROOF. We first define functions $h_{\alpha,\nu}$ ($\alpha \in \mathcal{A}, 1 \le \nu \le |\alpha|$) on M. Let K' be the maximal compact subgroup of $\mathcal{T}_{\Gamma'}$. Since the fundamental functions $\{h'_{\alpha,\nu}\}$ of $(M', \mathcal{F}')$ are K'-invariant, they are supposed to be defined on $M = U \times_{U(1)^r} M'$. So, we put

$$h_{\alpha,\nu} = \begin{cases} h'_{\alpha,\nu} & (\alpha \in \mathcal{A}') \\ \pi_\sharp^* h''_{\alpha,\nu} & (\alpha \in \mathcal{A}'') \end{cases}$$

Accordingly, we also put $c_{\alpha,\nu} = c'_{\alpha,\nu}$ if $\alpha \in \mathcal{A}'$ and $= c''_{\alpha,\nu}$ if $\alpha \in \mathcal{A}''$, where $c'_{\alpha,\nu}$ and $c''_{\alpha,\nu}$ are the fundamental constants.

We choose the ordering of $Y_{\alpha,0}, \dots, Y_{\alpha,|\alpha|}$ ($\alpha \in \mathcal{A}$) so that the ordering of $Y''_{\alpha,\nu} = (\pi_\sharp)_* Y_{\alpha,\nu}$ ($\alpha \in \mathcal{A}''$) and $Y'_{\alpha,\nu} = Y_{\alpha,\nu}$ ($\alpha \in \mathcal{A}'$) are equal to the ones induced from the fundamental functions $\{h''_{\alpha,\nu}\}$ and $\{h'_{\alpha,\nu}\}$ respectively (cf. the remark before Proposition 5.1). Hence the numbers $m_{\alpha,\nu}$ ($\alpha, \mathfrak{p}(\alpha)$, non-minimal) and $m_{\alpha,\nu} - m_{\alpha,0}$ ($\mathfrak{p}(\alpha)$, minimal) are uniquely determined. We define the number $m_{\alpha,0}$ for α such that $\mathfrak{p}(\alpha)$ is minimal by the formula (4.9) and the value of the constants $c_{\alpha,\nu}$. Then again by (4.9) the value of $e_{\beta\alpha}$ is determined for every non-maximal β and $\alpha \in \mathfrak{n}(\beta)$. It is also defined for all α and $\beta \prec \alpha$ so that Proposition 1.10 (2) is satisfied.

Now, let us define the function u_α on M by

$$u_\alpha = \prod_{\beta \prec \alpha} \prod_{\nu=1}^{|\beta|} (h_{\beta,\nu} + e_{\beta\alpha}).$$

Let D'' be the (horizontal) subbundle of TM given by Proposition 6.9. Since the Kähler form ω' on M' is supposed to be defined on M so that the kernel coincides with the horizontal subbundle D'', we can define a 2-form ω on M by

$$\omega = \omega' + \sum_{s=1}^{r} |u_{\alpha_s}| \pi_\sharp^* \omega''_s.$$

Also, we define $F_\alpha(\lambda)$ ($\alpha \in \mathcal{A}$) by the horizontal lift of the generating polynomial $F''_\alpha(\lambda)$ of $(M''\mathcal{F}'')$ if $\alpha \in \mathcal{A}''$, and by the formula (6.2) if $\alpha \in \mathcal{A}'$, where $\widetilde{E}''_s$ is the horizontal lift of the energy function E''_s of M''_s, and $F'_\alpha(\lambda)$ is the generating polynomial of $(M', \mathcal{F}')$. Let $\mathcal{F}$ be the vector space spanned by all the coefficients of $F_\alpha(\lambda)$ ($\alpha \in \mathcal{A}$).

Define the orthonormal frame $V_{\alpha,\nu}, IV_{\alpha,\nu}$ on the open $\mathcal{T}_\Gamma$-orbit M^1 by using the corresponding frames on M' and M'' in the obvious manner. Then we have the formula (6.1) and the relations

$$[W_{\alpha,\nu}, W_{\beta,\mu}] = [W_{\alpha,\nu}, IW_{\beta,\mu}] = [IW_{\alpha,\nu}, IW_{\beta,\mu}] = 0 \quad ((\alpha,\nu) \ne (\beta,\mu))$$

$$[V_{\alpha,\nu}, IV_{\alpha,\nu}] \equiv \operatorname{sgrad} a_{\alpha,\nu} \mod D_{\alpha,\nu},$$

where $W_{\alpha,\nu} = a_{\alpha,\nu}^{-1/2} V_{\alpha,\nu}$ and

$$a_{\alpha,\nu}^{-1} = |u_\alpha| \prod_{\mu \ne \nu} |h_{\alpha,\mu} - h_{\alpha,\nu}|.$$

Hence the arguments in Section 1 imply that ω is a Kähler form, and with this Kähler metric $(M, \mathcal{F})$ becomes a Kähler-Liouville manifold of type (A). The uniqueness and the property (2) obviously follow from the construction above. To prove (1) we need the following lemma.

LEMMA 6.12.

$$Y_{\alpha,\nu} = \widetilde{Y}''_{\alpha,\nu} + \frac{d''_{\alpha_s}}{d''_{\alpha,\nu}} \frac{v_\alpha(c_{\alpha,\nu})}{u_{\alpha_s}} Z_{\alpha_s} \qquad (\alpha \in \mathcal{A}''_s),$$

where $\widetilde{Y}''_{\alpha,\nu}$ *is the horizontal lift of* $Y''_{\alpha,\nu}$.

PROOF. Let θ be the connection form given by Proposition 6.9. Then we have

$$i_{Y_{\alpha,\nu}} d\theta = d''_{\alpha_s} \pi_\sharp^*(i_{Y''_{\alpha,\nu}} \omega''_s) \otimes Z_{\alpha_s} = -\frac{d''_{\alpha_s}}{d''_{\alpha,\nu}} d\left(\frac{v_\alpha(c_{\alpha,\nu})}{u_{\alpha_s}}\right) \otimes Z_{\alpha_s}.$$

Since the left-hand side is equal to $-d(\theta(Y_{\alpha,\nu}))$ by Proposition 6.9, we have

$$Y_{\alpha,\nu} = \widetilde{Y}''_{\alpha,\nu} + \frac{d''_{\alpha_s}}{d''_{\alpha,\nu}} \frac{v_\alpha(c_{\alpha,\nu})}{u_{\alpha_s}} Z_{\alpha_s} + (\text{constant term}).$$

Then, by comparing both sides at points on $L_{\alpha,\nu} = V(\mathbf{R}_{\geq 0} Y_{\alpha,\nu})$, the lemma is proved. □

The lemma above implies that $Y_{\alpha,\nu} = d_{\alpha,\nu}^{-1} \operatorname{sgrad} v_\alpha(c_{\alpha,\nu})$ for any $\alpha \in \mathcal{A}''_s$, where $d_{\alpha,\nu} = \epsilon(\alpha_s) d''_{\alpha,\nu}$ and $\epsilon(\alpha_s)$ is the sign of u_{α_s}. Since it is also true for $\alpha \in \mathcal{A}'$ ($d_{\alpha,\nu} = d'_{\alpha,\nu}$ in this case), the condition (1) is therefore satisfied. This completes the proof of Theorem 6.11. □

7. The case where $\#\mathcal{A} = 1$

In this section we shall classify compact Kähler-Liouville manifolds (of type (A)) such that $\#\mathcal{A} = 1$. Note that such manifolds are isomorphic to the complex projective space $\mathbf{C}P^n$ (with the standard $(\mathbf{C}^\times)^n = (\mathbf{C}^\times)^{n+1}/\mathbf{C}^\times$ action) as toric variety.

Let $(M, \mathcal{F})$ be a compact Kähler-Liouville manifold of type (A) such that the associated partially ordered set $\mathcal{A}$ consists of one element. In this case we write ν instead of (α, ν), and d_* instead of d_α. Put

$$S = \cap_{\nu=1}^{n-1} L_\nu, \quad \{q_1\} = L_0 \cap S, \quad \{q_2\} = L_n \cap S.$$

S is holomorphically isomorphic to $\mathbf{C}P^1$. We regard S as a Kähler manifold with the induced metric. Clearly, Y_n is tangent to S, and its zeros are q_1 and q_2. Let $\gamma(t)$ $(0 \leq t \leq l/2)$ be a minimal geodesic of unit speed such that $\gamma(0) = q_1$ and $\gamma(l/2) = q_2$. Since S is a surface of revolution, $\gamma(t)$ is extended to a closed geodesic of the least period l. Recalling the function $v(\lambda) = \prod_{\nu=1}^n (h_\nu - \lambda)$, put

$$h(t) = \frac{v(c_n)(\gamma(t))}{\prod_{\nu=0}^{n-1}(c_\nu - c_n)} \qquad t \in \mathbf{R}/l\mathbf{Z}.$$

PROPOSITION 7.1. $h \in C^\infty(\mathbf{R}/l\mathbf{Z})$ *possesses the following properties:*

(1) $h(-t) = h(t)$ *for any* t;
(2) $h(0) = 1$, $h(l/2) = 0$;
(3) $h'(t) < 0$ *if* $0 < t < l/2$;
(4) $h'(T_\nu) = -\sqrt{2d_* c_\nu(1 - c_\nu)}$ $(1 \leq \nu \leq n-1)$, *where* T_ν $(0 < T_\nu < l/2)$ *is defined by* $h(T_\nu) = c_\nu$;
(5) $-h''(0) = h''(l/2) = d_*$.

PROOF (EXCEPT (4)). (1), (2), and (5) are clear. Since $Y_n \neq 0$ at $\gamma(t)$ $(0 < t < l/2)$, (3) is also obvious. □

Let $\mathcal{C} = \mathcal{C}_n$ be the set of elements $(\{c_0, \dots, c_n\}, d_*, l, h)$ such that d_* and l are positive constants, $\{c_\nu\}$ are constants satisfying

$$1 = c_0 > c_1 > \cdots > c_n = 0, \tag{7.1}$$

and $h \in C^\infty(\boldsymbol{R}/l\boldsymbol{Z})$ satisfies the conditions (1), ..., (5) in Lemma 7.1. We say that two elements $(\{c_\nu\}, d_*, l, h)$ and $(\{\widetilde{c}_\nu\}, \widetilde{d}_*, \widetilde{l}, \widetilde{h})$ are *equivalent* if $\widetilde{d}_* = d_*$, $\widetilde{l} = l$ and either $\widetilde{c}_\nu = c_\nu$, $\widetilde{h} = h$, or

$$\widetilde{c}_\nu = 1 - c_{n-\nu}, \qquad \widetilde{h}(t) = 1 - h(\frac{l}{2} - t).$$

THEOREM 7.2. *The assignment of $(\{c_\nu\}, d_*, l, h) \in \mathcal{C}$ to $(M, \mathcal{F})$ described above gives the one-to-one correspondence between the set of the isomorphism classes of compact Kähler-Liouville manifolds of type* (A) *satisfying $\#\mathcal{A} = 1$ and the equivalence classes of elements of $\mathcal{C}$.*

To prove Proposition 7.1 (4) and Theorem 7.2 we shall use the results for (real) Liouville manifolds obtained in Part 1. First, we prove the following

PROPOSITION 7.3. *Let $p_0 \in M^1$ be the base point so that M is identified with the toric variety $X(\Delta)$. Put*

$$N = \mathrm{Exp}_{p_0}(D^+).$$

Then, N is a well-defined real submanifold of M, which is totally geodesic and diffeomorphic to $\boldsymbol{RP}^n$. Moreover, take the S^2TN-component F' of each element $F \in \mathcal{F}$ and put

$$\mathcal{F}' = \{F' \mid F \in \mathcal{F}\}.$$

Then $(N, \mathcal{F}')$ is a proper Liouville manifold of rank one and type (B), *and its core is isomorphic to*

$$(\boldsymbol{R}/l\boldsymbol{Z}, \{[h - c_1], \dots, [h - c_{n-1}]\}).$$

PROOF. Using the real number field $\boldsymbol{R}$ instead of $\boldsymbol{C}$ in the construction of the toric variety $X(\Delta)$ (cf. Section 5), one obtains a submanifold $X(\Delta)(\boldsymbol{R})$ diffeomorphic to the n-dimensional real projective space $\boldsymbol{RP}^n$. Since its tangent space is spanned by $I\mathfrak{k} = D^+$ at each point on $\boldsymbol{RP}^n \cap M^1$, Proposition 1.6 implies that $X(\Delta)(\boldsymbol{R})$ is totally geodesic and $N = X(\Delta)(\boldsymbol{R})$.

It is easy to verify that $(N, \mathcal{F}')$ is a Liouville manifold. Put

$$G_\nu = \sum_{\xi=1}^{n} \left(\prod_{\mu \neq \xi} (h_\mu - c_\nu) \right) (V_\xi^2 + (IV_\xi)^2) \in \mathcal{F} \qquad (1 \leq \nu \leq n-1).$$

Then we have

$$\{p \in N \mid (G'_\nu)_p = 0\} = M^s \cap L_\nu \cap N,$$
$$\{p \in N \mid \mathrm{rank}\,(G'_\nu)_p \leq 1\} = L_\nu \cap N.$$

Also, we have $(dG'_\nu)_\lambda \neq 0$ at some $\lambda \in T_p^*N$ for every $p \in M^s \cap L_\nu \cap N$, because $d(h_\nu + h_{\nu+1})$ does not vanish at p. Hence the Liouville manifold $(N, \mathcal{F}')$ is proper and of rank one. Since N is diffeomorphic to $\boldsymbol{RP}^n$, it is of type (B) (cf. Part 1, Theorems 3.3.1 and 3.4.1).

By definition the core of the Liouville manifold $(N, \mathcal{F}')$ consists of the 1 dimensional riemannian submanifold

$$C = \cap_{\nu=1}^{n-1}(M^s \cap L_\nu \cap N)$$

(called the core submanifold) and the equivalence classes $[\widetilde{h}_\nu]$ of the functions $\widetilde{h}_\nu$ on it defined by

$$(G'_\nu)_p = \widetilde{h}_\nu(p)V^2, \qquad p \in C$$

where V is the unit normal vector to $L_\nu \cap N$. Two functions (not identically zero) are said to be equivalent if the ratio is constant. It is easily seen that C is equal to the image of a closed geodesic passing through q_1 and q_2. Hence we can take γ so that its image coincides with C. Thus C is isometric to $\boldsymbol{R}/l\boldsymbol{Z}$, and $\widetilde{h}_\nu(\gamma(t)) = h(t) - c_\nu$. This completes the proof. □

Note that another choice of the base point p_0 gives another submanifold, but they are mutually transferred with the action of K. Hence the isomorphism class of the Liouville manifold $(N, \mathcal{F}')$ is uniquely determined. As was shown in Part 1, the isomorphism classes of proper Liouville manifolds of rank one are completely classified by means of the isomorphism classes of the cores. In the present case, two cores $(\boldsymbol{R}/l\boldsymbol{Z}, \{[h - c_1], \dots, [h - c_{n-1}]\})$ and $(\boldsymbol{R}/\widetilde{l}\boldsymbol{Z}, \{[\widetilde{h} - \widetilde{c}_1], \dots, [\widetilde{h} - \widetilde{c}_{n-1}]\})$ are isomorphic if and only if $l = \widetilde{l}$ and either $h(t) = \widetilde{h}(t)$, $c_\nu = \widetilde{c}_\nu$, or $h(t) = 1 - \widetilde{h}(-t + l/2)$, $c_\nu = 1 - \widetilde{c}_{n-\nu}$. Hence those isomorphism classes corresponds to the equivalence classes of elements of $\mathcal{C}$.

By the proof of Theorem 3.3.1 and Theorem 3.4.1 in Part 1, we obtain a branched covering of N whose covering space is a torus. We now explain it: Put

$$\alpha_\nu = 4\int_{T_{\nu-1}}^{T_\nu} \frac{dt}{\sqrt{(-1)^{\nu-1}\prod_{\mu=1}^{n-1}(h(t) - c_\mu)}} \qquad (1 \le \nu \le n), \tag{7.1}$$

where $T_\nu \in [0, l/2]$ is defined by $h(T_\nu) = c_\nu$ $(0 \le \nu \le n)$. Put

$$R = \prod_{\nu=1}^{n} (\boldsymbol{R}/\alpha_\nu \boldsymbol{Z}),$$

and let $x_\nu \pmod{\alpha_\nu \boldsymbol{Z}}$ be the natural coordinate of $\boldsymbol{R}/\alpha_\nu\boldsymbol{Z}$. Let $H(\simeq (\boldsymbol{Z}/2\boldsymbol{Z})^n)$ be the transformation group of R generated by $\tau_{2\nu} \circ \tau_{2\nu+1}$ $(1 \le \nu \le n-1)$ and $\tau_1 \circ \prod_{\nu=1}^{n} \tau_{2\nu}$, where

$$\begin{aligned} \tau_{2\nu-1}(x_1, \dots, x_n) &= (x_1, \dots, x_{\nu-1}, \frac{\alpha_\nu}{2} - x_\nu, x_{\nu+1}, \dots, x_n) \\ \tau_{2\nu}(x_1, \dots, x_n) &= (x_1, \dots, x_{\nu-1}, -x_\nu, x_{\nu+1}, \dots, x_n). \end{aligned} \tag{7.2}$$

Then we have

PROPOSITION 7.4. *There is a surjective mapping $\Phi : R \to N$ possessing the following properties:*

(1) *For any $p \in N$, $\Phi^{-1}(p)$ is an H-orbit;*
(2) $\Phi_*(\partial/\partial x_\nu) = \pm W_\nu$;
(3) *$h_\nu \circ \Phi$ are C^∞ functions;*
(4) $M^s \cap N = \{p \in N \mid \#\Phi^{-1}(p) < 2^n\}$;
(5) $L_\nu \cap N = \{\Phi(x) \mid x_\nu = 0, \alpha_\nu/2$ *or* $x_{\nu+1} = \pm\alpha_{\nu+1}/4\}$ $(1 \le \nu \le n-1)$, $L_0 \cap N = \{\Phi(x) \mid x_1 = \pm\frac{\alpha_1}{4}\}$, $L_n \cap N = \{\Phi(x) \mid x_n = 0, \frac{\alpha_n}{2}\}$;

(6) $\Phi \circ \tau_{2\nu-1} = \exp(\pi Y_{\nu-1}) \circ \Phi, \quad \Phi \circ \tau_{2\nu} = \exp(\pi Y_\nu) \circ \Phi.$

PROOF OF PROPOSITION 7.1 (4). Since the function $h_\nu \circ \Phi$ depends only on the variable x_ν, we write it $\widetilde{h}_\nu(x_\nu)$. Observing the formula

$$\operatorname{Hess} v(c_\nu) = d_* \prod_{\substack{0 \le \mu \le n \\ \mu \ne \nu}} (c_\mu - c_\nu)$$

at a point p such that $h_\nu(p) = c_\nu$ and $h_{\nu+1}(p) \ne c_\nu$, we have $\widetilde{h}'_\nu(0) = 0$ and

$$\widetilde{h}''_\nu(0) = (-1)^{n-\nu} d_* \prod_{\mu \ne \nu} (c_\mu - c_\nu).$$

Note that the vector fields V_ν and W_ν are locally well-defined (up to sign) and smooth as vector fields on $M^0 \cap N$, though they are not determined around $p \notin M^1$ as vector fields on M. Since $h(t) = h_\nu(\gamma(t))$ on $[T_{\nu-1}, T_\nu]$, we have

$$\begin{aligned}
h'(T_\nu)^2 &= \lim_{t \nearrow T_\nu} (V_\nu h_\nu)^2(\gamma(t)) \\
&= \lim_{t \nearrow T_\nu} \frac{(W_\nu h_\nu)^2(\gamma(t))}{(-1)^{n-\nu} \prod_{\mu \ne \nu} (h_\mu - h_\nu)} \\
&= \lim_{x_\nu \to 0} \frac{(\widetilde{h}'_\nu(x_\nu))^2}{(-1)^{n-\nu} \prod_{\mu=1}^{n-1} (c_\mu - \widetilde{h}_\nu(x_\nu))} \\
&= 2d_*(1 - c_\nu)c_\nu.
\end{aligned}$$

□

PROOF OF THEOREM 7.2. Let $(\{c_\nu\}, d_*, l, h) \in \mathcal{C}$ be an arbitrary element, and let $(N, \mathcal{F}')$ be a proper Liouville manifold of rank one whose core is isomorphic to

$$(\boldsymbol{R}/l\boldsymbol{Z}, \{[h - c_1], \ldots, [h - c_{n-1}]\}).$$

To prove Theorem 7.2 it suffices to show that there is a unique Kähler-Liouville manifold $(M, \mathcal{F})$ up to isomorphism such that the associated Liouville manifold is isomorphic to $(N, \mathcal{F}')$. To do so, we first review how to construct $(N, \mathcal{F}')$.

Let α_ν, R, $\tau_{2\nu-1}$, $\tau_{2\nu}$, and H be as above. It is not hard to see that R/H is homeomorphic to $\boldsymbol{RP}^n$ with the quotient topology. Put $N = R/H$, and let $\Phi : R \to N$ be the quotient mapping. To regard N as differentiable manifold diffeomorphic to $\boldsymbol{RP}^n$, it is necessary to specify coordinate systems around branch points (i.e., points $p \in N$ such that $\#\Phi^{-1}(p) < 2^n$). Let N^s denote the branch locus. Put

$$\begin{aligned}
I_\nu &= \{\Phi(x) \mid \tau_{2\nu}(x) = x \text{ and } \tau_{2\nu+1}(x) = x\} \quad (1 \le \nu \le n-1) \\
J_\nu &= \{\Phi(x) \mid \tau_{2\nu}(x) = x \text{ or } \tau_{2\nu+1}(x) = x\} \quad (0 \le \nu \le n)
\end{aligned}$$

Then $N^s = \cup_{\nu=1}^{n-1} I_\nu$. Let $p = \Phi(a) \in N^s$. Then there is a unique subset K of $\{1, \ldots, n-1\}$ such that $p \in I_\nu$ if and only if $\nu \in K$. Writing $K = \{\nu_1, \ldots, \nu_k\}$, $\nu_1 < \cdots < \nu_k$, we have $\nu_{i+1} - \nu_i \ge 2$. Define functions $y_1, \ldots, y_n$ by

$$\begin{aligned}
y_{\nu_i} &= (x_{\nu_i} - a_{\nu_i})^2 + (x_{\nu_i+1} - a_{\nu_i+1})^2 \quad (1 \le i \le k) \\
y_{\nu_i+1} &= 2(x_{\nu_i} - a_{\nu_i})(x_{\nu_i+1} - a_{\nu_i+1}) \quad (1 \le i \le k) \\
y_\nu &= x_\nu \qquad (\nu, \nu - 1 \notin K)
\end{aligned}$$

The system of functions (y_ν) is then projectable, and becomes a coordinate system around p mentioned above (cf. Part 1, Proposition 3.3.2).

Define C^∞ mappings

$$\boldsymbol{R}/\alpha_\nu \boldsymbol{Z} \to \begin{cases} [-T_1, T_1] & (\nu = 1) \\ [T_{\nu-1}, T_\nu] & (2 \le \nu \le n-1)\,, \\ [T_{n-1}, l - T_{n-1}] & (\nu = n) \end{cases} \qquad (x_\nu \mapsto t = t_\nu(x_\nu))$$

by the differential equations

$$t'_\nu(x_\nu)^2 = (-1)^{\nu-1} \prod_{\mu=1}^{n-1} (h(t_\nu) - c_\mu)$$

and the initial conditions

$$t_\nu(0) = T_\nu \quad (1 \le \nu \le n), \qquad \begin{cases} t'(0) = 0,\ t''_\nu(0) < 0 & (1 \le \nu \le n-1) \\ t'_n(0) < 0 & (\nu = n). \end{cases}$$

Put $\widetilde{h}_\nu(x_\nu) = h(t_\nu(x_\nu))$ and

$$g' = \sum_{\nu=1}^{n} (-1)^{n-\nu} \left(\prod_{\mu \ne \nu} (\widetilde{h}_\mu - \widetilde{h}_\nu) \right) (dx_\nu)^2$$

$$F'_\nu = \sum_{\mu=1}^{n} \frac{\prod_{\xi \ne \mu} (\widetilde{h}_\xi - c_\nu)}{(-1)^{n-\mu} \prod_{\xi \ne \mu} (\widetilde{h}_\xi - \widetilde{h}_\mu)} \left(\frac{\partial}{\partial x_\mu} \right)^2 \qquad (1 \le \nu \le n-1).$$

Then g' and F'_ν are projectable, and define the riemannian metric on N and the sections of S^2TN respectively. Denoting by E' the energy function associated with g' and by $\mathcal{F}'$ the vector space spanned by F'_ν $(1 \le \nu \le n-1)$ and E', we obtain the proper Liouville manifold $(N, \mathcal{F}')$ of rank one and type (B), whose core is isomorphic to the given one.

The functions $\widetilde{h}_\nu$ are also projectable, and define the continuous functions h_ν on N. The function h_ν is smooth outside $I_\nu \cup I_{\nu-1}$. Also, it is easily seen that the symmetric polynomials of $h_1, \dots, h_n$ are smooth on the whole N. Put $v(\lambda) = \prod_\nu (h_\nu - \lambda)$, and

$$X_\nu = \frac{1}{d_* \prod_{\substack{0 \le \mu \le n \\ \mu \ne \nu}} (c_\mu - c_\nu)} \operatorname{grad} v(c_\nu) \qquad (0 \le \nu \le n).$$

The following lemma is immediate.

LEMMA 7.5. (1) $[X_\mu, X_\nu] = 0$ *for any* μ, ν.
(2) $v(c_\nu)(p) = 0$, $(X_\nu)_p = 0$ *for* $p \in J_\nu$.
(3) $\operatorname{Hess} v(c_\nu)$ *on the normal bundle* NJ_ν *is equal to* $d_* \prod_{\mu \ne \nu} (c_\mu - c_\nu) g'$.

Let $\pi : \boldsymbol{R}^{n+1} - \{0\} \to \boldsymbol{RP}^n$ be the natural projection, and let $(w_0, \dots, w_n)$ be the natural coordinate system of $\boldsymbol{R}^{n+1}$.

PROPOSITION 7.6. *There is a diffeomorphism* $\phi : N \to \boldsymbol{RP}^n$ *such that*

$$\phi_*(X_\nu) = \pi_* \left(w_\nu \frac{\partial}{\partial w_\nu} \right) \qquad (0 \le \nu \le n).$$

PROOF. We first construct ϕ on $N - J_0$. Noting that $X_1, \dots, X_n$ are linearly independent at every point on $N^1 = N - \cup_{\nu=0}^{n} J_\nu$, we define (closed) 1-forms $\omega_1, \dots, \omega_n$ on N^1 by $\omega_\nu(X_\mu) = \delta_{\nu\mu}$. It is easily seen that ω_ν is smoothly extended to $N - (J_0 \cup J_\nu)$. Fix a point $p_0 \in N^1$

Let σ_ν $(0 \le \nu \le n)$ be the involution on N defined by $\Phi \circ \tau_{2\nu} = \sigma_\nu \circ \Phi$ or $\Phi \circ \tau_{2\nu+1} = \sigma_\nu \circ \Phi$. Then, σ_ν is the reflection with respect to J_ν, and preserves each element of $\mathcal{F}'$. Also, we see that $N - (J_0 \cup J_\nu)$ has two connected components; one contains p_0 and the other contains $\sigma_\nu(p_0)$.

Let x_ν $(1 \le \nu \le n)$ be the function on $N - (J_0 \cup J_\nu)$ defined by

$$x_\nu(p) = \begin{cases} \exp(\int_0^1 \omega_\nu(\dot{c}(t))dt) & (p \simeq p_0) \\ -\exp(\int_0^1 \omega_\nu(\dot{c}(t))dt) & (p \simeq \sigma_\nu(p_0)), \end{cases}$$

where $p \simeq p_0$ means p and p_0 are on the same component, and $c(t)$ $(0 \le t \le 1)$ is a curve in $N - (J_0 \cup J_\nu)$ from p_0 or $\sigma_\nu(p_0)$ to p. Clearly we have

$$x_\nu(\sigma_\nu(p)) = -x_\nu(p),$$
$$x_\nu(p_0) = 1, \quad x_\nu(\sigma_\nu(p_0)) = -1,$$
$$\omega_\nu = \frac{dx_\nu}{x_\nu}.$$

We now claim that x_ν is smoothly extended to $N - J_0$ by putting $x_\nu = 0$ on $J_\nu - J_0$. In fact it is an easy consequence of the following lemma.

LEMMA 7.7. *For each $p \in J_\nu - J_0$, there is a neighborhood U of p and a C^∞ function u_ν on U such that $u_\nu^2 = |v(c_\nu)|$.*

The lemma above follows from Lemma 7.5. Thus we have obtained the diffeomorphism

$$N - J_0 \to \boldsymbol{R}^n \quad (p \mapsto (x_1(p), \dots, x_n(p)), \tag{7.3}$$

which maps X_ν to $x_\nu \partial/\partial x_\nu$ and p_0 to $(1, \dots, 1)$ (the surjectivity follows from the completeness of X_ν on N^1). Now, making the coordinate functions on $N - J_\nu$ in the same way, and gluing them together, we consequently obtain the desired diffeomorphism $\phi : N \to \boldsymbol{RP}^n$. □

We now continue the proof of Theorem 7.2. By virtue of Proposition 7.6 we may identify N with $\boldsymbol{RP}^n$. Hence $X_\nu = \pi_*(\partial/\partial w_\nu)$, and J_ν is given by $w_\nu = 0$. Also, we regard $\boldsymbol{RP}^n$ as a submanifold of $\boldsymbol{CP}^n$ in the natural manner. The projection $\boldsymbol{C}^{n+1} - \{0\} \to \boldsymbol{CP}^n$ is also denoted by π. Let $K = U(1)^n$ be the torus acting on $\boldsymbol{CP}^n$ by

$$((\lambda_1, \dots, \lambda_n), \pi(w_0, \dots, w_n)) \to \pi(w_0, \lambda_1 w_1, \dots, \lambda_n w_n) \quad (|\lambda_\nu| = 1).$$

The following lemma is immediate.

LEMMA 7.8. *Let H be a symmetric 2-form on $\boldsymbol{RP}^n$ invariant with respect to σ_ν $(0 \le \nu \le n)$. Then there is a unique hermitian form $\widetilde{H}$ on $\boldsymbol{CP}^n$ satisfying the following conditions:*

(1) $\widetilde{H}|_{T\boldsymbol{RP}^n} = H$;

(2) $\widetilde{H}(X, IY) = 0$ *for any* $X, Y \in T_p\boldsymbol{RP}^n$, $p \in \boldsymbol{RP}^n$;

(3) $\widetilde{H}$ *is* K*-invariant.*

By the lemma above the riemannian metric g' extends to a hermitian metric g on $\boldsymbol{CP}^n$. Let $F' \in \mathcal{F}'$ be an arbitrary element. By using the bundle isomorphism $T\boldsymbol{RP}^n \to T^*\boldsymbol{RP}^n$ induced from g', F' is translated to a symmetric 2-form $F'_\flat$. Extending it to a hermitian form $F_\flat$ on $\boldsymbol{CP}^n$, and again translating with respect to g, we obtain a section F of $S^2T\boldsymbol{CP}^n$. Let $\mathcal{F}$ be the vector space (over $\boldsymbol{R}$) spanned by such F. Then direct calculations show that g is a Kähler metric, and with this metric $(\boldsymbol{CP}^n, \mathcal{F})$ becomes a Kähler-Liouville manifold of type (A) that satisfies $\#\mathcal{A} = 1$. This completes the proof of Theorem 7.2. □

8. Existence theorem

Let $(M, \mathcal{F})$ be a compact Kähler-Liouville manifold of type (A). For each $\alpha \in \mathcal{A}$, we define a Kähler-Liouville manifold $(M_\alpha, \mathcal{F}_\alpha)$ as follows: Define the closed subset $\mathcal{A}_\alpha$ of $\mathcal{A}$ by

$$\mathcal{A}_\alpha = \{\beta \in \mathcal{A} \mid \alpha \preceq \beta\}.$$

Let $(M'', \mathcal{F}'')$ be the Kähler-Liouville manifold that is the base space of the fibre bundle determined by the open subset $\mathcal{A} - \mathcal{A}_\alpha$. Next, regard $(M'', \mathcal{F}'')$ as the total space, and let $(M_\alpha, \mathcal{F}_\alpha)$ be the Kähler-Liouville manifold that is the typical fibre of the fibre bundle determined by the open subset $\{\alpha\}$ of $\mathcal{A}_\alpha$.

If $|\alpha| \geq 2$, then $(M_\alpha, \mathcal{F}_\alpha)$ is of type (A), and it possesses the structure of toric variety that is given by the structure of Kähler-Liouville manifold. In case $|\alpha| = 1$, we also regard M_α as a toric variety, whose structure is inherited from that of M''. So, in any case M_α is isomorphic to $\boldsymbol{CP}^{|\alpha|}$ as toric variety.

Let N be a 1-dimensional compact Kähler manifold which is also a toric variety such that the associated $U(1)$-action preserves the metric. We shall simply call it a compact toric Kähler manifold (of dimension 1). To such a manifold we assign positive constants d_*, l, and a function h on $\boldsymbol{R}/l\boldsymbol{Z}$ as follows: Let Y be an infinitesimal generator of the $U(1)$-action so that the least period of $\exp(sY)$ is 2π. The set of zeros of Y consists of two points, say q_0 and q_1. We may assume that the endomorphism ∇Y of $T_{q_1}N$ is equal to the complex structure I (then it is equal to $-I$ at q_0). Let $l/2$ be the distance between these two points. Then a minimal geodesic $\gamma(t)$ from q_0 to q_1 extends to a closed geodesic of least period l. Since the 1-form $i_Y\omega$ is closed (ω is the Kähler form), there is a unique function $\widetilde{h}$ on M such that

$$i_Y\omega = -d\widetilde{h}, \qquad \widetilde{h}(q_1) = 0.$$

Put $d_* = \widetilde{h}(q_0)^{-1}$ and $h(t) = d_*\widetilde{h}(\gamma(t))$.

The following lemma is immediate.

LEMMA 8.1. *(d_*, l, h) defined above possesses the following properties:*
(1) *$h(-t) = h(t)$ for any t;*
(2) *$h(0) = 1$, $h(l/2) = 0$;*
(3) *$h'(t) < 0$ if $0 < t < l/2$;*
(4) *$-h''(0) = h''(l/2) = d_*$.*

Let $\mathcal{C}_1$ be the set of (d_*, l, h) such that d_* and l are positive constants and h is a C^∞ function on $\boldsymbol{R}/l\boldsymbol{Z}$ satisfying the conditions $(1), \dots, (4)$ in Lemma 8.1. We say that two elements (d_*, l, h) and $(\widetilde{d}_*, \widetilde{l}, \widetilde{h})$ are equivalent if $d_* = \widetilde{d}_*$, $l = \widetilde{l}$, and either $h(t) = \widetilde{h}(t)$ or $h(t) = 1 - \widetilde{h}(l/2 - t)$. For consistency with the definition of $\mathcal{C}_n$, we shall also write $(\{1, 0\}, d_*, l, h)$ instead of (d_*, l, h).

LEMMA 8.2. *The assignment above gives the one-to-one correspondence between the set of the isomorphism classes of 1-dimensional compact toric Kähler manifolds and the set of the equivalence classes of elements of* $\mathcal{C}_1$.

The proof is easy. We also have the following

PROPOSITION 8.3. *Theorem 6.11 is also valid under the assumption that* M' *is a one-dimensional compact toric Kähler manifold.*

The Proof is similar to that of Theorem 6.11, so we shall omit. Now, we state the main theorem in this section, which will imply the existence of compact Kähler-Liouville manifold of type (A) whose structure of toric variety is prescribed. Let M be a toric variety of KL-A type, and let $\mathcal{A}$ be the associated partially ordered set. Let $m_{\alpha,\nu}$ ($\alpha \in \mathcal{A}$, not minimal, $0 \le \nu \le |\mathsf{p}(\alpha)|$) be numbers satisfying Proposition 4.21 with which M is defined. Let $c_{\alpha,\nu}$ ($0 \le \nu \le |\alpha|$, $\alpha \in \mathcal{A}$), $e_{\beta\alpha}$ ($\beta \prec \alpha$), and d_α ($\alpha \in \mathcal{A}$) be constants that satisfy the conditions (4.1), (4.2), (4.8), and (4.9). In this case we say that the constants $\{c_{\alpha,\nu}, e_{\beta\alpha}, d_\alpha\}$ are *compatible* with the toric variety M (of KL-A type). Note that the compatibility has no meaning in case $\#\mathcal{A} = 1$.

REMARK. 1. M determines only the differences $m_{\alpha,\nu} - m_{\alpha,0}$ for α such that $\mathsf{p}(\alpha)$ is minimal. Hence for such α one can choose $m_{\alpha,0}$ arbitrary so that they satisfy Proposition 4.21 (5).

2. $\{m_{\alpha,\nu}\}$ just determine every $e_{\beta\alpha}$, every ratio $d_{\mathsf{p}(\alpha)}/d_\alpha$, and $\{c_{\alpha,\nu}\}$ for every non-maximal α. Hence one can choose $d_\alpha > 0$ arbitrary for minimal α, and also $c_{\alpha,\nu}$ arbitrary for maximal α so that they satisfy (4.1).

THEOREM 8.4. *Let* M *be a compact toric variety of KL-A type, and let* $\mathcal{A}$ *be the associated partially ordered set. Let* $\{c_{\alpha,\nu}, e_{\beta\alpha}, d_\alpha\}$ *be constants compatible with* M. *For each* $\alpha \in \mathcal{A}$, *choose* $l_\alpha > 0$ *and* $h_\alpha \in C^\infty(\boldsymbol{R}/l_\alpha \boldsymbol{Z})$ *so that* $(\{c_{\alpha,\nu}\}, |d_\alpha|, l_\alpha, h_\alpha) \in \mathcal{C}_{|\alpha|}$. *Then there is a unique structure of Kähler-Liouville manifold* $(M, \mathcal{F})$ *of type* (A) *over the toric variety* M *possessing the following properties:*

(1) *The associated structure of toric variety is identical with the given one;*
(2) *the fundamental constants, the conjunction constants, and the scaling constants are equal to* $\{c_{\alpha,\nu}\}$, $\{e_{\beta\alpha}\}$, *and* $\{d_\alpha\}$ *respectively;*
(3) *for each* $\alpha \in \mathcal{A}$, *the induced Kähler-Liouville manifold (the toric Kähler manifold if* $|\alpha| = 1$*)* $(M_\alpha, \mathcal{F}_\alpha)$ *corresponds to the equivalence class represented by the given element*

$$(\{c_{\alpha,\nu}\}, |d_\alpha|, l_\alpha, h_\alpha) \in \mathcal{C}_{|\alpha|}.$$

PROOF. We prove this theorem by induction on $\#\mathcal{A}$. The case $\#\mathcal{A} = 1$ follows from Theorem 7.2. Let $k \ge 2$, and assume that the theorem is true for the case where the number of elements of the associated partially ordered set is less than k.

Now, let M and $\mathcal{A}$ be as above, and suppose $\#\mathcal{A} = k$. We may assume that $\mathcal{A}$ is connected. Let $\alpha_0 \in \mathcal{A}$ be the minimal element, and put $\mathcal{A}' = \{\alpha_0\}$, $\mathcal{A}'' = \mathcal{A} - \mathcal{A}'$. As before, let $\mathcal{A}'' = \cup_{s=1}^r \mathcal{A}''_s$ be the decomposition into connected components, and α_s the minimal element of $\mathcal{A}''_s$. Let M' and M'' be the associated toric varieties. As is easily seen, the constants $c_{\alpha,\nu}$ ($\alpha \in \mathcal{A}''$), $e_{\beta\alpha}$ ($\beta \in \mathcal{A}'', \beta \prec \alpha$), $\epsilon(\alpha_s) d_\alpha$ ($\alpha \in \mathcal{A}''_s, 1 \le s \le r$) are compatible with the toric variety M'' of KL-A type, where $\epsilon(\alpha_s)$ is the sign of d_{α_s}. So, by induction assumption we obtain a unique structure of Kähler-Liouville manifold $(M'', \mathcal{F}'')$ over the toric variety M'' possessing the properties stated in the theorem.

Also, by Theorem 7.2 and Lemma 8.2 there is a unique structure of Kähler-Liouville manifold (or toric Kähler manifold if $|\alpha_0| = 1$) $(M', \mathcal{F}')$ over the toric variety M' corresponding to the element

$$(\{c_{\alpha_0,\nu}\}, d_{\alpha_0}, l_{\alpha_0}, h_{\alpha_0}) \in \mathcal{C}_{|\alpha_0|}.$$

Then, by Theorem 6.11 we obtain a structure of Kähler-Liouville manifold $(M, \mathcal{F})$ over the toric variety M such that $(M', \mathcal{F}')$ and $(M'', \mathcal{F}'')$ are isomorphic to the ones induced from $(M, \mathcal{F})$. It is clear that $(M, \mathcal{F})$ possesses the properties (1) and (2). (3) follows from the fact that the Kähler-Liouville manifold (or the toric Kähler manifold) $(M_\alpha, \mathcal{F}_\alpha)$ $(\alpha \in \mathcal{A}'')$ induced from $(M, \mathcal{F})$ is isomorphic to the one induced from $(M'', \mathcal{F}'')$.

This fact also proves the uniqueness of $(M, \mathcal{F})$. In fact, let $(\widetilde{M}, \widetilde{\mathcal{F}})$ be another Kähler-Liouville manifold possessing the properties stated in the theorem, and let $(\widetilde{M'}, \widetilde{\mathcal{F}'})$ and $(\widetilde{M''}, \widetilde{\mathcal{F}''})$ be the induced ones. Then the fact mentioned above and the induction assumption indicate that $(\widetilde{M''}, \widetilde{\mathcal{F}''})$ is isomorphic to $(M'', \mathcal{F}'')$, and Theorem 7.2 and Lemma 8.2 indicate that $(\widetilde{M'}, \widetilde{\mathcal{F}'})$ is isomorphic to $(M', \mathcal{F}')$. Hence by Theorem 6.11, $(\widetilde{M}, \widetilde{\mathcal{F}})$ is isomorphic to $(M, \mathcal{F})$. This completes the proof. $\square$

References

[Be] A. Besse, *Manifolds all of whose geodesics are closed*, Springer-Verlag, 1978.

[Br] A. V. Brailov, *Construction of completely integrable geodesic flows on compact symmetric spaces*, Izv. Akad. Nauk SSSR **50** (1986), 661–674; English transl. in Math. USSR Izv. **29** (1987).

[Da] V. I. Danilov, *The geometry of toric varieties*, Russian Math. Surveys **33** (1978), 97–154.

[De] M. Demazure, *Sous-groupes algébriques de rang maximum du groupe de Cremona*, Ann. Sci. École Normal Sup. **3** (1970), 507–588.

[Fo] A. T. Fomenko, *Symplectic geometry*, Advanced Studies in Contemporary Mathematics, Gordon and Breach Science Publishers, 1988.

[Fu] W. Fulton, *Introduction to toric varieties*, Annals of Math. Studies 131, Princeton University Press, 1993.

[I] M. Igarashi, *On Kähler-Liouville surfaces*, to appear in J. Math. Soc. Japan.

[IKS] M. Igarashi, K. Kiyohara, and K.Sugahara, *Noncompact Liouville surfaces*, J. Math. Soc. Japan **45** (1993), 459–479.

[IW] K. Ii and S. Watanabe, *Complete integrability of the geodesic flows on symmetric spaces*, Advanced Studies in Pure Math. 3, 1984, pp. 105–124.

[Ki1] K. Kiyohara, *Compact Liouville surfaces*, J. Math. Soc. Japan **43** (1991), 555–591.

[Ki2] ———, *On infinitesimal $C_{2\pi}$-deformations of standard metrics on spheres*, Hokkaido Math. J. **13** (1984), 151–231.

[Ki3] ———, *Riemannian metrics with periodic geodesic flows on projective spaces*, Japan. J. Math. **13** (1987), 209–234.

[Kl1] W. Klingenberg, *Lectures on closed geodesics*, Springer, Berlin-Heidelberg, 1978.

[Kl2] ———, *Riemannian Geometry*, 2nd ed., Walter de Gruyter, Berlin-New York, 1995.

[Kob1] S. Kobayashi, *Differential geometry of complex vector bundles*, Publications of the Mathematical Society of Japan 15, Iwanami Shoten and Princeton University Press, 1987.

[Kob2] ———, *Transformation groups in differential geometry*, Springer-Verlag, Berlin Heidelberg New York, 1972.

[Kol1] V. N. Kolokol'tsov, *Geodesic flows on two-dimensional manifolds with an additional first integral polynomial in velocities*, Izv. Akad. Nauk. SSSR **46** (1982), 994–1010; English transl. in Math. USSR Izv. **21** (1983).

[Kol2] ———, *New examples of manifolds all of whose geodesics are closed*, Vestnik Mosk. Gos. Univers. (1984), 80–82. (Russian)

[L] J. Lützen, *JOSEPH LIOUVILLE 1809–1882*, Studies in the History of Mathematics and Physical Sciences 15, Springer-Verlag, Berlin-Heidelberg, 1990.

[Mi] A. S. Mishchenko, *Integration of geodesic flows on symmetric spaces*, Mat. Zametki **31** (1982), 257–262; English transl. in Math. Notes **31** (1982).

[MF1] A. S. Mishchenko and A. T. Fomenko, *Generalized Liouville's method of integration of Hamiltonian systems*, Funktsional. Anal. i Prilozhen. **12** (1978), 46–56; English transl. in Functional Anal. Appl. **12** (1978).

[MF2] A. S. Mishchenko and A. T. Fomenko, *Euler equations on finite-dimensional Lie groups*, Izv. Akad. Nauk SSSR **42** (1978), 396–415; English transl. in Math. USSR Izv. **12** (1978).

[Mo] J. Moser, *Integrable Hamiltonian Systems and Spectral Theory*, Lezioni Fermiane, Accademia Nazionale Dei Lincei, Pisa, 1981.

[O1] T. Oda, *Convex bodies and algebraic geometry*, Springer-Verlag, Berlin Heidelberg, 1988.

[O2] ———, *Lectures on torus embeddings and applications (Based on joint work with Katsuya Miyake)*, Tata Inst. of Fund. Research 58, Springer-Verlag, Berlin Heidelberg New York, 1978.

[Th] A. Thimm, *Integrable geodesic flows on homogeneous spaces*, Ergodic Theory and Dynamical Systems **1** (1981), 495-517.

[Ts1] C. Tsukamoto, *Integrability of Infinitesimal Zoll Deformations*, Advanced Studies in Pure Mathematics 3, Geometry of Geodesics and Related Topics, Kinokuniya and North-Holland Publishing, 1984, pp. 97–104.

[Ts2] ———, *Infinitesimal Blaschke conjectures on projective spaces*, Ann. Sci. École Norm. Sup. **14** (1981), 339–356.

[V] H. Viesel, *Die Gestalt analytischer Liouvillescher Flächen im Großen*, Math. Ann. **166** (1966), 175–186.

[Y] K. Yamato, *A class of Riemannian manifolds with integrable geodesic flows*, J. Math. Soc. Japan **47** (1995), 719–733.

Editorial Information

To be published in the *Memoirs*, a paper must be correct, new, nontrivial, and significant. Further, it must be well written and of interest to a substantial number of mathematicians. Piecemeal results, such as an inconclusive step toward an unproved major theorem or a minor variation on a known result, are in general not acceptable for publication. *Transactions* Editors shall solicit and encourage publication of worthy papers. Papers appearing in *Memoirs* are generally longer than those appearing in *Transactions* with which it shares an editorial committee.

As of July 31, 1997, the backlog for this journal was approximately 8 volumes. This estimate is the result of dividing the number of manuscripts for this journal in the Providence office that have not yet gone to the printer on the above date by the average number of monographs per volume over the previous twelve months, reduced by the number of issues published in four months (the time necessary for preparing an issue for the printer). (There are 6 volumes per year, each containing at least 4 numbers.)

A Copyright Transfer Agreement is required before a paper will be published in this journal. By submitting a paper to this journal, authors certify that the manuscript has not been submitted to nor is it under consideration for publication by another journal, conference proceedings, or similar publication.

Information for Authors and Editors

Memoirs are printed by photo-offset from camera copy fully prepared by the author. This means that the finished book will look exactly like the copy submitted.

The paper must contain a *descriptive title* and an *abstract* that summarizes the article in language suitable for workers in the general field (algebra, analysis, etc.). The *descriptive title* should be short, but informative; useless or vague phrases such as "some remarks about" or "concerning" should be avoided. The *abstract* should be at least one complete sentence, and at most 300 words. Included with the footnotes to the paper, there should be the 1991 *Mathematics Subject Classification* representing the primary and secondary subjects of the article. This may be followed by a list of *key words and phrases* describing the subject matter of the article and taken from it. A list of the numbers may be found in the annual index of *Mathematical Reviews*, published with the December issue starting in 1990, as well as from the electronic service e-MATH [**telnet e-MATH.ams.org** (or **telnet 130.44.1.100**). Login and password are **e-math**]. For journal abbreviations used in bibliographies, see the list of serials in the latest *Mathematical Reviews* annual index. When the manuscript is submitted, authors should supply the editor with electronic addresses if available. These will be printed after the postal address at the end of each article.

Electronically prepared papers. The AMS encourages submission of electronically prepared papers in AMS-TEX or AMS-LATEX. The Society has prepared author packages for each AMS publication. Author packages include instructions for preparing electronic papers, the *AMS Author Handbook*, samples, and a style file that generates the particular design specifications of that publication series for both AMS-TEX and AMS-LATEX.

Authors with FTP access may retrieve an author package from the Society's Internet node `e-MATH.ams.org` (130.44.1.100). For those without FTP access, the author package can be obtained free of charge by sending e-mail to `pub@ams.org` (Internet) or from the Publication Division, American Mathematical Society, P.O. Box 6248, Providence, RI 02940-6248. When requesting an author package, please specify $\mathcal{A}\mathcal{M}\mathcal{S}$-TEX or $\mathcal{A}\mathcal{M}\mathcal{S}$-LATEX, Macintosh or IBM (3.5) format, and the publication in which your paper will appear. Please be sure to include your complete mailing address.

Submission of electronic files. At the time of submission, the source file(s) should be sent to the Providence office (this includes any TEX source file, any graphics files, and the DVI or PostScript file).

Before sending the source file, be sure you have proofread your paper carefully. The files you send must be the EXACT files used to generate the proof copy that was accepted for publication. For all publications, authors are required to send a printed copy of their paper, which exactly matches the copy approved for publication, along with any graphics that will appear in the paper.

TEX files may be submitted by email, FTP, or on diskette. The DVI file(s) and PostScript files should be submitted only by FTP or on diskette unless they are encoded properly to submit through e-mail. (DVI files are binary and PostScript files tend to be very large.)

Files sent by electronic mail should be addressed to the Internet address `pub-submit@ams.org`. The subject line of the message should include the publication code to identify it as a Memoir. TEX source files, DVI files, and PostScript files can be transferred over the Internet by FTP to the Internet node `e-math.ams.org` (130.44.1.100).

Electronic graphics. Figures may be submitted to the AMS in an electronic format. The AMS recommends that graphics created electronically be saved in Encapsulated PostScript (EPS) format. This includes graphics originated via a graphics application as well as scanned photographs or other computer-generated images.

If the graphics package used does not support EPS output, the graphics file should be saved in one of the standard graphics formats—such as TIFF, PICT, GIF, etc.—rather than in an application-dependent format. Graphics files submitted in an application-dependent format are not likely to be used. No matter what method was used to produce the graphic, it is necessary to provide a paper copy to the AMS.

Authors using graphics packages for the creation of electronic art should also avoid the use of any lines thinner than 0.5 points in width. Many graphics packages allow the user to specify a "hairline" for a very thin line. Hairlines often look acceptable when proofed on a typical laser printer. However, when produced on a high-resolution laser imagesetter, hairlines become nearly invisible and will be lost entirely in the final printing process.

Screens should be set to values between 15% and 85%. Screens which fall outside of this range are too light or too dark to print correctly.

Any inquiries concerning a paper that has been accepted for publication should be sent directly to the Editorial Department, American Mathematical Society, P. O. Box 6248, Providence, RI 02940-6248.

Selected Titles in This Series

(Continued from the front of this publication)

590 **Chris Jantzen,** Degenerate principal series for symplectic and odd-orthogonal groups, 1996

589 **James Damon,** Higher multiplicities and almost free divisors and complete intersections, 1996

588 **Dihua Jiang,** Degree 16 Standard L-function of $GSp(2) \times GSp(2)$, 1996

587 **Stéphane Jaffard and Yves Meyer,** Wavelet methods for pointwise regularity and local oscillations of functions, 1996

586 **Siegfried Echterhoff,** Crossed products with continuous trace, 1996

585 **Gilles Pisier,** The operator Hilbert space OH, complex interpolation and tensor norms, 1996

584 **Wayne W. Barrett, Charles R. Johnson, and Raphael Loewy,** The real positive definite completion problem: Cycle completability, 1996

583 **Jin Nakagawa,** Orders of a quartic field, 1996

582 **Darryl McCollough and Andy Miller,** Symmetric automorphisms of free products, 1996

581 **Martin U. Schmidt,** Integrable systems and Riemann surfaces of infinite genus, 1996

580 **Martin W. Liebeck and Gary M. Seitz,** Reductive subgroups of exceptional algebraic groups, 1996

579 **Samuel Kaplan,** Lebesgue theory in the bidual of $C(X)$, 1996

578 **Ale Jan Homburg,** Global aspects of homoclinic bifurcations of vector fields, 1996

577 **Freddy Dumortier and Robert Roussarie,** Canard cycles and center manifolds, 1996

576 **Grahame Bennett,** Factorizing the classical inequalities, 1996

575 **Dieter Heppel, Idun Reiten, and Sverre O. Smalø,** Tilting in Abelian categories and quasitilted algebras, 1996

574 **Michael Field,** Symmetry breaking for compact Lie groups, 1996

573 **Wayne Aitken,** An arithmetic Riemann-Roch theorem for singular arithmetic surfaces, 1996

572 **Ole H. Hald and Joyce R. McLaughlin,** Inverse nodal problems: Finding the potential from nodal lines, 1996

571 **Henry L. Kurland,** Intersection pairings on Conley indices, 1996

570 **Bernold Fiedler and Jürgen Scheurle,** Discretization of homoclinic orbits, rapid forcing and "invisible" chaos, 1996

569 **Eldar Straume,** Compact connected Lie transformation groups on spheres with low cohomogeneity, I, 1996

568 **Raúl E. Curto and Lawrence A. Fialkow,** Solution of the truncated complex moment problem for flat data, 1996

567 **Ran Levi,** On finite groups and homotopy theory, 1995

566 **Neil Robertson, Paul Seymour, and Robin Thomas,** Excluding infinite clique minors, 1995

565 **Huaxin Lin and N. Christopher Phillips,** Classification of direct limits of even Cuntz-circle algebras, 1995

564 **Wensheng Liu and Héctor J. Sussmann,** Shortest paths for sub-Riemannian metrics on rank-two distributions, 1995

563 **Fritz Gesztesy and Roman Svirsky,** (m)KdV solitons on the background of quasi-periodic finite-gap solutions, 1995

562 **John Lindsay Orr,** Triangular algebras and ideals of nest algebras, 1995

561 **Jane Gilman,** Two-generator discrete subgroups of $PSL(2, R)$, 1995

560 **F. Tomi and A. J. Tromba,** The index theorem for minimal surfaces of higher genus, 1995

559 **Paul S. Muhly and Baruch Solel,** Hilbert modules over operator algebras, 1995

558 **R. Gordon, A. J. Power, and Ross Street,** Coherence for tricategories, 1995

557 **Kenji Matsuki,** Weyl groups and birational transformations among minimal models, 1995

(See the AMS catalog for earlier titles)